MONITORING THE STANDARDS OF EDUCATION

PAPERS IN HONOR OF JOHN P. KEEVES

Titles of Related Interest

ANDERSON, RYAN & SHAPIRO
The IEA Classroom Environment Study

BURSTEIN
The IEA Study of Mathematics III: Student Growth and Classroom Processes

ELLEY
The IEA Study of Reading Literacy: Achievement and Instruction in Thirty-Two School Systems

GORMAN, PURVES & DEGENHART
The IEA Study of Written Composition I: The International Writing Tasks and Scoring Scales

HUSÉN, TUIJNMAN & HALLS
Schooling in Modern European Society: A Report of the *Academia Europaea*

KEEVES
The IEA Study of Science III: Changes in Science Education and Achievement: 1970 to 1984

PELGRUM & PLOMP
The IEA Study of Computers in Education: Implementation of an Innovation in 21 Education Systems

ROSIER & KEEVES
The IEA Study of Science I: Science Education and Curricula in Twenty-Three Countries

TRAVERS & WESTBURY
The IEA Study of Mathematics I: Analysis of Mathematics Curricula

MONITORING THE STANDARDS OF EDUCATION

PAPERS IN HONOR OF JOHN P. KEEVES

Edited by

Albert C. Tuijnman

Organization for Economic Cooperation and Development,
Center for Educational Research and Innovation,
France

and

T. Neville Postlethwaite

University of Hamburg, Institute of Comparative Education,
Germany

PERGAMON

U.K.	Elsevier Science Ltd, The Boulevard, Langford Lane, Kidlington, Oxford OX5 1GB, U.K.
U.S.A.	Elsevier Science Inc., 660 White Plains Road, Tarrytown, New York 10591-5153, U.S.A.
JAPAN	Elsevier Science Japan, Tsunashima Building Annex, 3-20-12 Yushima, Bunkyo-ku, Tokyo 113, Japan

First edition 1994

Library of Congress Cataloging in Publication Data

Monitoring the standards of education: papers in honor of John P. Keeves / edited by Albert C. Tuijnman, T. Neville Postlethwaite. – 1st ed.
p. cm.
Includes bibliographical references and index.
1. Education–Standards–Evaluation. 2. Educational evaluation. 3. Educational tests and measurements. 4. Educational indicators. 5. Keeves, John P.
I. Tuijnman, Albert. II. Postlethwaite, T. Neville.
LB3060.82.M66 1994
379.1'58 – dc20
94-15706

British Library Cataloguing in Publication Data

A catalogue record for this book is available from the British Library

ISBN 0 08 042386 8

Printed and bound in Great Britain by Redwood Books, Trowbridge

Sponsored by

The International Academy of Education

The general aim of the Academy is to foster scholarly excellence in all fields of education. Towards this end, the Academy's goals are:

− To create an international network of scholars to write state-of-the-art reports on major educational issues, to establish permanent relations among relevant disciplines of education, and to identify excellent practices wherever they might be found;

− To disseminate knowledge about effective policies and practices to interested educators;

− To conduct advanced training, particularly seminars for education officials, research workers, and other key staff;

− To provide critical and evaluative perspectives on studies and issues in the forefront of educational debate;

− To identify research priorities relating to critical issues in education;

− To strengthen communication and cooperation between educational researchers and educational practitioners.

Contents

List of Figures

List of Tables

Preface

It was in the 19th century that many countries introduced compulsory primary schooling and education became a key element in nation-state building. Ever since those days, ministries of national education have set norms for the adequacy of school buildings, the equipment and supplies of schools, teacher qualifications, and certain procedures in school, including in many cases a prescribed syllabus. Often, these regulations were binding, and adherence to them was strictly enforced by the inspectorate. Corrective action was taken if schools and classrooms failed to adhere to the norms.

Traditionally, the norms set by education ministries concerned the level of inputs into the school system: how many teachers for how many students, what type of facilities, how much money? These norms were intended to guarantee that the schools would be equipped properly so that all students would have adequate opportunities to learn. In many countries the authorities went one step further by specifying the content to be learned, and that, in primary school all students should be exposed to a similar curriculum. In secondary school, there were syllabi for the different types of schools. It was common for the curriculum frameworks in many countries to be prescribed in minute detail.

Since the 1960s and 1970s, however, the situation has been gradually changing in a number of ways. First, the ministries of education in many countries have sought to encourage flexibility and accountability in the school system by loosening, to some extent, their hold over input, opportunity, and even content regulations. Rather than focusing on the input norms, many countries have increasingly come to direct attention to the specification of goals for education and the setting of performance standards for students, teachers, schools and, indeed, the system as a whole.

Before meaningful performance standards can be set, it must be decided what the students should know and be able to do. A decision about what students must learn and master presupposes a degree of consensus on the goals, objectives, and purposes of schooling. Once goals are agreed and standards set, a further step is to produce evidence about the extent to which they are being met. This implies the collection of data for performance assessment and the monitoring of educational progress. But, any analysis of the data is incomplete and can yield spurious results if information about the contexts, inputs, and processes of schooling is not taken into account. The interpretation of assessment data thus necessitates the collection and analysis of information in several domains and at several levels in the system.

The emphasis on outcomes presents a radical departure from the traditional approach to the collection of data for resource management and educational planning. It implies new roles for all actors—the students, teachers and school principals, the central and local school administrations, the inspectorate, the agencies responsible for the collection of statistical data about the school system, and educational planners and decision-makers. The purpose of this book is to describe the changes and their ramifications. It is addressed mainly to educational planners and managers, university teachers and students, and others interested in the monitoring of educational progress and, more particularly, the setting and assessment of performance standards.

The first five chapters in this volume deal with the background and overall aspects of why monitoring is conducted, how it can be done, the purposes and organization of education indicator systems, the role of international comparisons in monitoring, what performance standards mean, and what approaches to standard setting exist. These chapters are mainly descriptive. The next seven chapters deal with the methodology of standard setting and performance assessment, and are more technical in their orientation. They explain issues in the monitoring of cognitive and affective outcomes, address the concepts of validity and reliability, classify the various procedures and methods for the setting of cut-scores for standards, and discuss their advantages and disadvantages, depending on the circumstances and purposes for which the standards are to be used. These chapters also offer many insights into how performance standards are set and used in the assessment of educational progress in the United States, a country where there has been monitoring for nearly three decades, and where the issue of standard setting is now at the forefront of the debate.

ATn & TNP
February, 1994

John P. Keeves

A Tribute to John P. Keeves

John Philip Keeves—or JPK as he came to be known—was born in South Australia on September 20, 1924. The editors of this *Festschrift* encountered John only late in his career. Neville Postlethwaite met him in 1964 when John attended his first meeting at the International Association for the Evaluation of Educational Achievement (IEA), and Albert Tuijnman met him in 1987 when he was at the Institute of International Education, Stockholm University. We took it upon ourselves to plan this book because we felt that it would offer an appropriate means of marking the seventieth birthday of a man who has devoted not only his professional career but indeed a major part of his life to the development of educational research. We shall first describe his career, then his scholarly work, and finally say a few things about John as a person.

Life Career

JPK has been a school teacher, a university lecturer, an educational researcher, a research director, a writer and editor, and a committee man. He has also been, and still is, a professorial fellow. JPK took his first university degrees—including a first-class honors degree in physics—at the University of Adelaide. From 1947 to 1949 he was a physics teacher at Prince Alfred College in Adelaide—the same school he himself had attended as a pupil from 1934 to 1942. To gain other experience he then taught physics at Radley College in England, where some of his pupils are now famous members of Parliament. While in England, he took a diploma in education in 1951 and a Master of Philosophy degree at Oxford University in 1952. Having accomplished this he returned to Prince Alfred College and remained there as a science teacher until 1962. During this decade he wrote several reports on mathematics, science, and education. He also became an active member of the Australian Science Teachers Association, an organisation of which he later became Honorary Federal Secretary. While at Prince Alfred College he coached the (Australian rules) football team and also played cricket. For sports connoisseurs, one of the football players coached by JPK was Ian Chappell, who later became the first of three Chappell brothers to play on the Australian cricket team.

In 1962 he moved to Melbourne to take up a position as senior research officer at the Australian Council for Educational Research (ACER). While at the ACER he studied on a part-time basis. In 1962-63, he won the Cohen Prize and took a Bachelor of Education degree (1st class honors) at the University of Melbourne. Later, in 1966, he obtained a Master of Education degree—with theoretical and applied statistics as separate subjects—at the same university. At the ACER, he was, interalia, in charge of the Australian component of the First International Mathematics Study of IEA. It was at this juncture in his life that he first experienced the sometimes uneasy relationship between researchers and policymakers. The first research report he wrote at the ACER—the report on the Australian part of the IEA study—was shredded because its empirical findings contradicted a policy just promulgated by one of the State Directors of Education.

From 1967 to 1971, JPK received leave of absence from the ACER to undertake doctoral studies in sociology at the Australian National University.

He collected data in the Canberra school system and wrote a dissertation, entitled *The Home, the School, and Educational Achievement: A multivariate study of the contributions of the home, the school, and the peer-group to change in mathematics and science performance during the first year at secondary school.* He used the same data set to write a second thesis, *Educational Environment and Student Achievement,* for which he was awarded a Ph.D. in educational psychology at Stockholm University, where Torsten Husén was his 'Doktorvater'.

He returned to the ACER as Associate Director in 1972 and became its Director in 1977, a position he held until he turned 60 in 1984. However, unlike others who retire at this age, JPK sought new challenges. He became visiting professor at the University of Melbourne (1985-86) and professorial fellow at Stockholm University (1987-89), before returning to his home city of Adelaide to take up a fellowship at the Flinders University of South Australia (1990-present).

During these years JPK was also a committee man. From 1977 to 1981 he was vice-Chair of the Council of the Institute of Early Childhood Development in Victoria. His interest in this field had been stimulated by his mother, who was a kindergarten teacher and an executive on the local school council. During this period he was also a member of the Committee on Education of the Australian National Committee for UNESCO and the ACT Schools Accrediting Agency. In 1980 he was appointed Chair of the Committee of Enquiry into Education in South Australia. Further to this he served on various IEA committees.

Scholarship

Most of JPK's work has been concerned with the identification of factors that account for differences in achievement between students, between schools, and between the education systems of different countries. This has required not only a knowledge of the appropriate research methodologies available at any one time but also a thorough understanding of the sociological, psychological, instructional and school-organizational variables that influence student learning. Although JPK has conducted studies on student achievement in several subjects, notably science, mathematics, literacy and numeracy, his major interest has been in explaining differences in science achievement.

He began his IEA association in the first mathematics study but, in 1967, became the driving force on the steering committee of the First International Science Study (FISS), a committee chaired by Sam Comber. Indeed, JPK spent nearly a year in Stockholm in 1971-72 working on the data analyses and write up of the science report, *Science Education in Nineteen Countries,* a study that became known throughout the world as the Comber/Keeves report. When IEA began its Second International Science Study (SISS) in 1980, JPK was the obvious choice to chair the steering committee. Because funds were insufficient, many delays were encountered at the data analysis and reporting stage. To see the job through until the end, and despite the inconvenience this move caused to himself, JPK agreed to go to Stockholm for two years (1987-89) to complete the work. The product spans two volumes: *Science Education and Curricula in Twenty-three Countries* (with Malcolm Rosier) and *Changes in Science Education and Achievement: 1970 to 1984.*

JPK has used his training in sociology extensively in his research work. He has conducted a number of studies on the measurement of home background factors. Other original studies have examined the role of student motivation and aspiration as variables mediating the effects of home background on learning and achievement. Another example of work in this field is his publication, *Schooling and Society in Australia* (with Larry Saha), which is still in use as a textbook. Methodologically, he has devoted much effort to the building of causal path models and, since the mid-1980s, to the development of statistical models for examining multi-level effects among variables at the student, class, school, and system level (see Keeves et al., *International Journal of Educational Research 14*, 3). He has also devoted energy to the problem of scaling achievement data between different levels of a school system and over time. Occasionally, he has ventured into curricular matters and the philosophy of science, but the main thrust has been on empirical research studies. This quantitative orientation is clearly present in the many entries he wrote and edited for the two editions of the *International Encyclopedia of Education*. These entries were also published separately in a monumental, 143-article volume edited by JPK, entitled: *Educational Research, Methodology, and Measurement: An International Handbook*. All in all, he has written—either alone or in association with others—14 books and over 100 journal articles.

JPK's sociological views have been much influenced by the work of Anthony Giddens. Although some researchers regard JPK as conservative in the theories and models he develops, those who have worked closely with him have a different view. For them, JPK is best characterized by his passion for being at the cutting edge of research methodology and statistical data analysis. He is curious about new theoretical approaches and eager to examine the latest analytical procedures. At the same time, JPK has a healthy, critical attitude both to new developments in theory and research methodology and to his own approach.

Personal Annecdotes

All of the above is a rather dry account of JPK's life career and scholarship; it has not dealt with him as a person. There are many stories about how generous he is with his time in teaching and supervising students, and also about how cautious he is with his money.

John P. Keeves is a determined man. He works at problems until they are solved—whatever the sacrifice. In directing a research effort, it is said that he forms a work matrix in his head, with him being the only one to know how all the bits fit together. He then assigns different persons to work on the various cells. If work on one cell gets out of kilter with the others, he is not pleased and can be very stubborn. However, by the end of his tenure as the Director of ACER he had managed, in a way, to unite a staff of 90 persons!

It has often been said that JPK's idea of paradise is to work 12 hours a day for seven days a week. Most weeks he undoubtedly achieved this 'ideal' state. It has further been said that he cannot understand why others do not do the same. Neville Postlethwaite once landed at Melbourne airport at 6 a.m. on a Sunday morning, after having flown all the way from Germany. In the car on the way from the airport to JPK's home, Neville asked JPK if he could have the next day off to recuperate and begin work on the Tuesday morning. "Fine", said JPK. But halfway on that ride JPK said that he had to check into the office for a while. It was 5 p.m. when they emerged from the ACER. And, on Monday it was a working day for both of them!

Typically, JPK showed up at the office at 7.15 a.m. each day (Mondays to Fridays) and left at about 6 p.m. He still took work home each evening. He also took work to his Summer house at Aldinga Beach in South Australia, where his study commands a panoramic view of St. Vincent's Gulf. Here he continues to write in the evenings, taking breaks only to read in the Agatha Christie novels he collects or to watch brief TV programs such as 'Dad's Army', 'Dinner for One', or 'Rumpole of the Bailey'. At Aldinga Beach, JPK starts and ends each day with long walks along the ocean front. If visitors are present they are taken along, and very often the discussion will be about work.

There are many persons in various countries of the world—from Brazil to Lesotho to Sweden and Finland, to the Netherlands, Germany, the US and Australia—who are in his debt for the amount of work and patience he invested in teaching and helping them with their statistical analyses and the writing of their dissertations. In 1988 and 1989, when Albert Tuijnman was a doctoral student at the Institute of International Education at Stockholm University, JPK voluntarily acted as resource person and supervised students. He was very generous with his advice, and could be relied upon to read draft chapters and scrutinize computer print out in a few days rather than weeks. Through his coaching, he transferred some of his fascination with statistical model building, and those who know his work can easily see his influence in the path models that characterize most of his students' dissertations. In 1992, he was given the 'Supervisor of the Year Award' at the Flinders University of South Australia. Had there been a prize of that kind at Stockholm University, JPK would no doubt have won that too!

Even as we write, JPK continues to work on the problem of measuring change over time. He is also studying the problem of how performance standards in education can best be set and measured. John Philip Keeves has high standards of work, and he expects high standards from others. We dedicate this book to him, and wish him many more years of the kind of life he wants for himself.

T. Neville Postlethwaite & Albert C. Tuijnman
(aided and abetted by Larry Saha & Petra Lietz)

Bordeaux & Paris, 1994.

Chapter 1

Monitoring Standards in Education: Why and How it Came About*

TORSTEN HUSÉN† and ALBERT TUIJNMAN‡

† Institute of International Education, Stockholm University, Sweden
‡ Center for Educational Research and Innovation, Organization for Economic Cooperation and Development, Paris, France

This opening chapter presents working definitions of key concepts and paints, with sweeping strokes, a picture of the historical background to the current endeavors to evaluate and monitor national systems of education. This overview further offers a number of insights into the forces, political, economic and others, that gave voice to the three main questions considered in this chapter—why the need for setting standards burgeoned; who were the ones asking for standards to be set, both nationally and internationally, and what the policy implications and practical consequences of standard setting might be.

There is no doubt that the monitoring of education systems has become a major policy issue in the post-Sputnik era. One of the most formidable tasks facing the international research community in education since the late 1950s has been to establish cross-nationally valid standards of curriculum content and student performance. Quite evidently, the interest in assessing the standards achieved in education and, not least, the readiness to monitor them continously have—with ups and downs—been gradually rising and made a political breakthrough in several countries since the mid-1980s. There are some central issues connected with the setting of standards and initiating monitoring at the

---

*Note: The views expressed in this chapter are those of the authors. They do not in any way commit the Organization for Economic Cooperation and Development or its Member countries.

national level. They can be expressed by means of the following questions:

(1)   Why should standards be assessed and monitoring activities undertaken?
(2)   Who determines the standards, and who are the "customers"?
(3)   What are the policy implications and the practical consequences of standard setting and monitoring?

The aim of this chapter is not to try to develop exhaustive, universally acceptable answers to these questions. But by referring to trends and historical examples, the chapter hopefully presents an elucidating perspective on why and how the assessment of educational progress and the evaluation of education systems came about, and how evaluation was brought into focus by monitoring systems, i.e., by following up, over time, how these systems work and what they "produce".

This chapter is divided into three main sections. The first offers working definitions of the key concepts dealt with in this volume. The second presents an historical overview of developments and phases in the conduct of educational evaluation and monitoring from the late 1950s to the 1990s. Conclusions and tentative answers to the questions raised above are offered in the third and last section.

Concepts and their Definitions

Standards, goals, assessment, monitoring, evaluation and accountability are major concepts frequently used in this volume. It is in order to try to give at least preliminary definitions of them before embarking on the task to sketch their role in education since the 1950s.

A *standard* refers to the degree of excellence required for particular purposes, a measure of what is adequate, a socially and practically desired level of performance (cf. Livingston, 1985). An education standard can be described as the specification of a desired level of content mastery and performance. Three types of education standards are considered in this volume: opportunity to learn standards, content standards, and performance standards. Briefly, the argument about standards goes as follows (National Academy of Education, 1993): for meaningful and fair performance standards to be set, it is necessary to define the exact content areas to which these standards shall apply. Before performance can be fairly assessed, it is moreover necessary to determine whether all the students have had adequate opportunities to learn the prescribed content.

Content defines the various areas of knowledge and skill which all the students should learn; *content standards* define those parts of the content in the curriculum that *all* the students should master; and *performance standards* describe how well the students should perform in these content areas. The chapters in this volume deal mostly with performance standards. It will be seen that there is no firm agreement on what such standards are, how they are best set, and what their relationship is or should be to the content items used in scoring student performance. What is clear, though, is that content standards and performance standards are interdependent, and that they are conditioned by *opportunity to learn* criteria.

Conceptually, an education standard can be seen as the specification of outcomes associated with specific *education goals*, which are often set at the national level by the central government. Goals for education are usually couched in very general terms and are not directly amenable to measurement. Hence they must be translated into more tangible objectives, formulated either as growth targets for learning or as curriculum benchmarks to be attained by a certain group of students and by a certain time. Thus, an educational standard is a fixed measure against which learning progress and achievement in particular areas of the curriculum can be judged. A physical standard—e.g., length, or weight—can be objectively defined and is invariant over time. The main problem of standard setting in education is that the underlying goals are subjectively derived and subject to change with the passing of time. An educational standard can be expressed as the mean score of a well-defined student population on a set of test items that measure how well the students have mastered knowledge elements in a certain area of the curriculum at a certain time.

Assessment refers to the techniques used in collecting information about educational outcomes either subjectively by using experienced judgments or by means of standardized, objective tests measuring cognitive—but sometimes also noncognitive—aspects of learning and student performance. It follows that there are different approaches to assessment, each serving particular purposes and implying a certain style of measurement. Chapter 3 explains the two main approaches that are commonly distinguished—norm-referenced and criterion-referenced assessment.

Monitoring commonly refers to systematic and regular procedures for the collection of data about important aspects of education at national, regional or local levels. A monitoring activity usually involves the collection of assessment data, but is not necessarily restricted to outcome variables. A coherent approach to the monitoring of educational progress must also take account of contextual information and measures of resource inputs and the processes of education.

A common definition of *evaluation* is the one given by Beeby (1977), who describes it as "the systematic collection and interpretation of evidence, leading, as part of the process, to a judgment of value with a view to action". Evaluation thus involves the making of a *value judgment* on the basis of evidence obtained through the measurement of attributes, characteristics and phenomena, whereby outcomes are related to certain goals and/or values set for the educational activities.

Although monitoring and evaluation are obviously related, there is also an important difference. Monitoring involves the systematic collection of evidence about the contexts, inputs, processes and outcomes of an education system. Evaluation involves the act of using the collected data for making a value judgment about the situation. The systematic collection of evidence about educational performance, as in an indicator system for the monitoring of educational progress, is an important element of evaluation in a model of *accountability*. Monitoring refers to ways in which accountability is ensured by using the evaluative judgment for purposes of influence in a managerial or other control system.

Conducting Educational Evaluation and Monitoring

Educational Evaluation: The Early Days

A major effort in evaluation and monitoring was launched in the 1930s by the Progressive Education Association in the United States wanting to contribute to a fundamental overhaul of American secondary education. Thirty schools were selected for an experiment with a new curriculum and unorthodox teaching strategies. An evaluation and follow-up staff was formed under the leadership of Ralph W. Tyler, then at the University of Chicago. Apart from measuring how well the students performed cognitively, they were also followed with regard to affective characteristics, such as social attitudes and critical thinking. A major focus was on how students having graduated from the experimental schools fared when they went through college in comparison with students leaving the traditional schools (Aikin, 1942). How did the efforts in the experimental schools to train independent study skills, cooperation and critical thinking pay off? Were their students handicapped with respect to basic rather than applied knowledge? What about the higher mental processes? The evaluation staff had to develop a number of new instruments in order to measure attitudes and other affective qualities, which meant an important step forward in the development of psycho-metric instruments.

After all, the Eight Year Study with its 30 schools was a rather limited although pioneering effort to study the effects of an educational, or rather didactic, reform in comparison with more recent attempts to monitor a wide range of standards at the regional or national levels of education systems. Nor had the idea of conducting cross-national evaluations yet been contemplated (Bloom et al., 1970). But it spearheaded the evaluation movement, and Tyler spelled out the *raison d'être* of evaluation in education in which he emphasized the importance of a coherent approach all the way from curriculum planning to the assessment of student performance in key subjects and cross-curriculum skills, such as critical thinking (Tyler, 1989).

Until the mid-1950s evaluation was a close to unknown concept among European educators and few attempts were made to conduct large-scale evaluation studies. One exception was research conducted from about 1952 to 1959 by the Swedish National Board of Education where comparisons of outcomes were made between the new comprehensive pilot schools and traditional schools at the lower secondary level (cf. Svensson, 1962). Evaluation began to be discussed in Europe when large-scale educational reforms were being planned and—in some instances—actually launched. Consequently a need to develop a methodological toolbox was felt, and educational researchers increasingly began to examine and draw upon the American experience in the art of educational evaluation (Husén, 1986).

The UNESCO Institute in Hamburg, Germany, founded in 1952, played an instrumental role in promoting evaluation in Europe. In the mid-1950s it started a series of conferences that provided a platform where educational researchers mainly from Europe and North America could exchange experience in educational evaluation and measurement and share knowledge about research efforts on issues such as school marks, grade-repeating, and the assessment of student performance. This initiated interesting debates about the possibility of comparing student achievement cross-nationally in different areas of the school curriculum.

This discussion took on new meaning, both academically and politically, in the Fall of 1957, when the Soviet Union succeeded in putting its first Sputnik into orbit. The repercussions of this event, not least in the United States, were great. It was regarded as an indication of Soviet superiority in space technology but also in the last run as a reflection of a superior mathematics and science education in Soviet schools. It gave critics in the United States fuel to attack the standards of performance in American schools, their curricula and teaching methods, particularly in high schools, a criticism that had flared up already in the early 1950s (Bestor, 1953). Improving the curricula and methods of teaching in mathematics and science became priorities. The federal government and Congress, which according to the Constitution were not allowed to support public schools—which was the prerogative of the states and local authorities—had to find a formula which legalized such a support. The formula was inherent in the title of the National Defense Education Act of 1958. This marked the entry into a new era in framing education policy at the national level in the United States, and later in Europe, mainly under the auspices of the Paris-based Organization for Economic Cooperation and Development (OECD).

That same year another significant event occurred. In 1958, at a meeting at the UNESCO Institute for Education in Hamburg, C. Arnold Anderson, Director of the newly established Center for Comparative Education at the University of Chicago proposed that one should try to conduct strict comparisons between national systems of education using empirical measures of resource inputs and student achievement. What he had in mind was to conduct a cross-national evaluation of the cognitive "yield" of various education systems by means of achievement tests administered to comparable samples of students. This idea was further discussed at a seminar held at Eltham Palace in Britain that same year. Leading US experts in educational measurement and testing as well as in curriculum development, such as Robert Thorndike of Teachers College and Benjamin Bloom of Chicago, a student and longstanding coworker of Ralph Tyler, were among those attending the meetings. These meetings led to the launching of a feasibility study conducted from 1959 to 1961 in a dozen countries, all of them more or less developed. Available test items from various subject areas, such as mathematics, science and reading comprehension, were translated into the eight languages involved (and back-translated independently) and administered to judgment samples of 13-year-olds in all the 12 countries. How it began and was conducted has been reported more in detail by Foshay (1962) and Husén (1967, Vol. 1, pp. 25 ff).

The 1959-61 pilot study was conducted in order to provide answers to—among others—the following questions: Would it be feasible to develop a uniform methodology for student testing that could be used in conducting cross-national assessments? Would it be possible to process and analyze the data thus collected uniformly and to make meaningful, cross-national comparisons? The feasibility study offered some encouragement and led in 1961 to the establishment of the International Association for the Evaluation of Educational Achievement (IEA), its first large-scale international study of mathematics achievement (Husén, 1967), and its subsequent legal incorporation under Belgian law in 1966.

Evaluation and Assessment in the 1960s

In 1960 Theodore Schultz (1961) gave his famous presidential address on education and economic growth at the annual meeting of the American Economic Association. American social scientists had developed a range of survey techniques, not least methods for conducting sample surveys that made it possible to draw inferences from samples to target populations. Psychometricians in the United States and the United Kingdom had developed group testing techniques and the instruments and methods needed for measurement and data analysis. The concept of a system and its logical extension, systems analysis, or the concept of human capital and its application in manpower planning are examples of innovations in sociological and economic thought that—together with advances in psychometrics and psychological research more generally—made it possible in the early 1960s to embark on the development of what was to become an empirical approach in comparative education.

The new concept of human capital provided legitimacy for the rapid expansion of systems of education during the 1960s. The main tenet, which supported by educational planners has hold to this time in not least the industrialized nations, was that the general capacity of the school system is a major determinant of educating a competent, productive and competitive work force and, consequentially, a major determinant of a nation's prosperity. Early in the 1950s, Japan's *Monbusho* (Ministry of Education) had developed a national plan according to which the education system would be instrumental in increasing the gross income per capita by one hundred per cent during the 1960s, something that was achieved with a broad margin.

Those times were characterized by optimism about what the social sciences, including research in education, could achieve in offering a basis for decision-making, and reform planning to improve society. There also was widespread support for systems theory and the application of linear models in social science research. Important was the work conducted on social indicators, which had its roots in the pioneering efforts to measure social change by William O'Brien and his colleagues at the University of Chicago in the 1940s and 1950s (Wyatt, 1994). The work on social indicators was spurred by a research program launched in 1962 by the American Academy of Arts and Sciences that sought to measure the second-order effects of the space exploration program, and followed on the apparent success of economists in influencing policy decisions by means of aggregate indicators constructed on the basis of national accounts statistics. The work on social indicators was extended to education in the mid-1960s, and was taken up by international organizations, such as UNESCO and the OECD (cf. Bauer, 1966). Two volumes reporting the results of the first IEA study of mathematics achievement in 12 countries (Husén, 1967) were published at this junction in time, as was the seminal publication, *Equality of Educational Opportunity* by James S. Coleman and his associates (1966).

Thus the stage was set for a new era of comparisons of standards achieved in various national systems of education. But, still, the survey techniques contemplated were cross-sectional, not longitudinal. Among those involved, the interest was limited to questions about the relative level of student performance achieved in various systems at a particular point in time. Still there was, during the 1960s, little or no talk about the measurement of age or grade-related learning curves in education systems, not to speak of the regular monitoring of education standards in a comparative perspective and in relation to contextual information and data on resource inputs. IEA held a meeting at Lake Mohonk in 1967 during which a group of social scientists tried to develop an input-output model of the process of educational achievement within the framework of national economies (Super, 1969). The United States excepted, most countries supported only a rather limited data collection effort in education. Interest was not with the assessment of the student, system and labor market outcomes of education, but with the keeping of administrative records useful for budget accounting and the management of inputs into education. The main question asked was how many schools, classrooms and teachers would be needed for how many students?

In the early 1960s, long before a similar interest surfaced in Europe and elsewhere, the then US Commissioner of Education, Francis Keppel, asked the Carnegie Corporation of New York to appoint an exploratory committee on assessing the *progress* of education. The committee, chaired by Ralph Tyler, the undisputed father of evaluation in education, had to come up with a plan for a system that would provide periodical information about the educational achievement of young people in the country. So far no dependable data on the level of student competence in the country as a whole had been available. Under Tyler, who also was then the Director of the Center for Advanced Study in the Behavioral Sciences at Stanford, the work of the committee progressed and eventually led to the first National Assessment of Educational Progress (NAEP), funded by the federal government and conducted in 1969.

Jumping ahead in history, in 1978 the US Congress enacted legislation requiring *periodic* national assessments of educational progress to be undertaken, which thus became institutionalized. In the late 1960s the NAEP was sold as a means of obtaining an "educational GNP", the educational equivalent of a macro-economic indicator, the implication being that it had to be repeated periodically, i.e. to be part of a monitoring system. However, for more than a decade no comparisons were made between the states—the monitoring of progress had to encompass the whole country and its major regions—but in the mid-1980s the chief state school officers recommended that comparisons be made between the states. The NAEP has been reporting the results of congressionally-mandated surveys of the academic achievement of American students on a state-by-state basis since 1990. In that year, 37 states, the District of Columbia, and two territories participated in a trial state assessment program in eight-grade mathematics. Voluntary participation increased to 41 states in 1992. The next step came a year later when the US Department of Education (1993) published a study, *Education in States and Nations,* comparing 16 .input and performance indicators calculated for the US states and 20 OECD countries, and presenting the results in tables mixing both kinds of entities. The report not only allowed state-to-state and country-to-country comparisons, but state-to-country comparisons as well.

Evaluation and Assessment in the 1970s

During the 1960s, education policymakers in the industrialized countries had been concerned mainly with the restructuring of selective secondary education and managing the rapid expansion of enrollments. These objectives were a response to an increase in the social demand for education. Satisfying this demand was the principal goal of education policy, and the flow of resources to the education sector grew with the expansion of the intake capacity of the education systems. The surplus created by rapid economic growth had sustained expansion and accommodated the rising costs of education caused by growing enrollments and increased staffing. But the first oil crisis in the early 1970s and the ensuing years of austerity put this to an end, as the fiscal resources needed to sustain the growth of systems of education became less abundant. The impact of the austerity in public budgets and other factors, such as the effects on education of a decline in the birth rate and the waning of optimism about what could be achieved by means of a massive investment in education—for example the idea that education could make a society both more wealthy and more equal—had a strong influence on education policy and, consequently on the agendas for educational research. In the research arena, controversy over systems theory and its linear models of reality in many circles stifled the belief in a fact-finding mission of the social sciences.

In April 1973, when the initial enthusiasm for social and education indicators had already begun to wane, the OECD published a document entitled "A framework for educational indicators to guide government decisions" (OECD, 1973). The 46 indicators described in the study were the result of a five-year program and were intended as measures of the effects of education on the individual and the society. Although the set of indicators was endorsed at the political level by the OECD Education Committee, they were never implemented. The era of an almost naive belief in what social and educational research could contribute in improving education and, more generally, providing a scientific and rational basis for the planning of a modern, industrial society had already begun to wane. Support for systems theory and the explanatory approach in the social sciences diminished, and critical questions were increasingly asked about the desirability of continuing the support for large-scale data collection.

In the United States the NAEP continued, albeit at a slower pace and at a lower level of ambition. It became increasingly difficult to sustain stakeholder interest in assessment and hence find the funds needed to undertake large-scale comparative studies. The IEA had published the findings from its famous six-subject study already in the period from 1973 to 1976 (Walker, 1976). It was to be the last such survey to be undertaken that decade. By the mid-1970s, macro-level educational planning had become highly suspect; instead the capacity of education systems to resist structural and procedural reform was emphasized. At UNESCO and the OECD, the ambitious agenda, established in the 1960s, for coordinated international data collection in education was abandoned shortly after the International Standard Classification of Education (ISCED) was implemented in the mid-1970s. This classification, developed as a means for aggregating national data categories into an international system, was criticized for rigidity and the lack of context information. Since a country's education system is deeply rooted in its history and part and parcel of its socio-cultural matrix, a classification system that lacks a capacity for including contextual information useful for interpreting the results is of limited use. This argument, while sensible, undermined faith in the validity of international comparisons of education systems, and was effectively used to diminish the scope of attempts to develop cross-national standards for student achievement in certain subject areas.

Specificity became a new catchword in educational research, and a host of innovative, qualitative studies yielded findings that were interpreted as lending support for the argument that, since each local school situation is considered unique, aggregate comparisons are meaningless and even harmful. Whatever research interest there was in the functioning of education as a system mostly disappeared, and micro-based studies concerned with teachers and questions of innovation at the school and classroom level became increasingly the mainstream. A price has been paid for this weakening of the basis for empirical studies in comparative education research, although it should be noted that, on the whole and on the positive side of the balance, the stocktaking exercise did provide a partly new agenda for research and gave a major impetus to the development of new research methods.

Assessment and Monitoring in the 1980s

The second oil crisis and other structural problems in the late 1970s and early 1980s inspired a quest for cost-effectiveness in education. As the objective increasingly became one of bringing the rapidly rising costs of education under control, attention shifted from a policy of satisfying social demand to those of the efficient management of resources and the effectiveness of educational provision. The focus on cost-effectiveness in turn called attention to issues of student learning and the *quality* of education. In the United States, concern with the standards of education was further raised by arguments presented in the seminal report, entitled *A Nation at Risk: The Imperative For Educational Reform* (US National Commission on Excellence in Education, 1983). There was no hiding the harsh criticism of American education the report advanced. To make matters worse, inadequate and low quality schooling was proclaimed a threat to national prosperity and security:

> If an unfriendly foreign power had attempted to impose on America the mediocre educational performance that exists today, we might well have viewed it as an act of war. As it stands, we have allowed this to happen to ourselves. We have even squandered the gains in student achievement made in the wake of the Sputnik challenge. ... We have, in effect, been committing an act of unthinking, unilateral educational disarmament. (p. 5)

The *A Nation at Risk* report played a pivotal role in raising concern over the quality of education. In a more subtle way it also raised political awareness that international comparative studies of student achievement could be used for the benchmarking of performance levels in different education systems. IEA had published the results of the second international mathematics study just before the report came out. Although a host of data was collected and revealing analyses of relationships among predictive variables and outcomes were produced (Robitaille & Garden, 1989), media attention focused mainly on the comparisons of the mean performance of 13-year-old students in arithmetic, algebra, and geometry. The results showed Japan in a leading and the United States, along with a few other Western European countries, in a low-achievement position. These for the United States unfavorable results underscored the main points advanced in the

A Nation at Risk report, and served to focus the attention of influential stakeholders on the use of measures of student achievement as criteria for judging the performance of the education system.

When the Ministers of Education of the OECD countries met in Paris in 1984 the shifting of priorities from issues in managing the quantitative growth of the education system to cost-effectiveness and quality assurance was striking. Questions of accountability—to what extent do the "customers" of education get "value for money"—were asked. This went hand in hand with doubts raised about the efficiency of the public sector. Thus the stage for school policy was set for the rest of the 1980s. In several European countries and the United States the result was that market or quasi-market approaches were contemplated—albeit rarely implemented in education policy—and that certain decision-making capabilities were devolved from the national or state level to local authorities and schools. However, other aspects of education policy were increasingly brought into the realm of national decision-making.

These trends led to another significant development in the 1980s. Until that time, many countries had not bothered to specify concrete goals as well as content and performance standards for school education. If goals were stated at all, they were formulated in the context of the nation-state, and they were phrased in a way that did not require comparisons against some national or international standard. They often included general statements about the purposes of education, and perhaps indicated how education should be organized so as to promote equal access. Content standards existed at the national level in some countries, notably Finland, France, Greece, Italy, Japan, New Zealand, and Quebec province in Canada. These curriculum standards were employed to ensure that all students have equal access to educational opportunities. But few OECD countries had specified concrete standards for student performance in primary and secondary education.

Following the chain of events beginning in the United States in the mid-1980s, the emerging trend has been not only to examine the goals and put them on paper but also to formulate the goals in terms of targets or performance standards to be achieved. Japan, New Zealand and the United Kingdom were early in specifying objectives for student achievement and identifying procedures for student assessment. In several countries—not least the United States—this trend was reinforced by the publication of the results of IEA's second international science study (IEA, 1988; Postlethwaite & Wiley, 1992; Keeves, 1992).

In September 1989 President Bush brought the governors of the 50 states together in Charlottesville, Virginia, in order to frame national goals for American education, an initiative which previously had been regarded as close to "unconstitutional". In the late 1930s the National Education Association, the leading teacher association in the United States, had set up an Educational Policies Commission which was an entirely private endeavour. In contrast, the six over-riding education goals proclaimed in 1990 were stated at the federal level, and they formed part of a comprehensive and long-term strategy for coping with the problems of American education (US Department of Education, 1991). One of the goals states that American youth will perform at the top of the competency ladder in mathematics and science by the year 2000. This goal reflects the thinking, widespread in the United States, that if excellence is to be achieved, then high standards must be set, and performance must be judged against those standards. So, once the six national education goals were in place, a National Education Goals Panel was appointed in July 1990. By reporting on the progress toward achieving world-class standards in education, its main purpose is to hold the nation and the states accountable for their attainment (US National Education Goals Panel, 1991-1993).

The specification of the six national education goals in the United States made it necessary to reflect on the appropriateness of education, and hence on the standards for learning that would define the outcomes to be expected from the system. This challenge was handed to the National Council on Education Standards and Testing (NCEST), established in June 1991. An unprecedented movement to develop nationwide education standards was initiated once the US Congress had endorsed, in January 1992, the main standard setting strategy as proposed in the report, *Raising Standards for American Education* (NCEST, 1992; US White House, 1993).

Ambitions of this kind have served to enhance the interest of several governments—no doubt that of the United States—to support international surveys of student achievement. The changes in policy have also influenced the research agenda. For example, studies of "what works" became popular from about the mid-1980s onwards and the "effective schools" movement grew rapidly. These developments combined to rekindle an interest in system-wide studies of education. Support grew for a theoretical framework in which attention was paid not only to "conventional" measures of the inputs into education—mainly money and people—but also, significantly, to process aspects and a wider array of outcomes of education. This new, system-wide framework influenced thinking about the nature of educational evaluation. Even though large-scale evaluation studies had been conducted in many countries since the 1960s, most initiatives concerned *ad hoc* studies limited in scope to the evaluation of particular innovations and experiments. By contrast, on both sides of the Atlantic, during the 1980s the entire education system became the subject of planned, systemic evaluation in a framework for the monitoring of educational progress. Thus the aim was no longer a modest one—for example, identifying the possible short-comings of certain school programs and implementing counter-measures—but was enlarged to embrace a demand for information about the functioning and effectiveness of the system as a whole.

So, gradually, *systemic* approaches to assessment and evaluation in a national monitoring framework emerged in a number of countries. As new information needs became apparent, the time-honored model of information gathering by central statistical offices no longer sufficed. The old, administrative approach to education statistics, which had relied mainly on registers with expenditure data and counts of teachers and students at different levels in the national education system, was supplemented by new, survey-based approaches to data collection.

By the early 1990s coherent approaches to the monitoring of educational progress were in place in an increasing number of countries, among which Australia (Victoria), Canada (British Columbia and Quebec), France, Ireland, Singapore, Sweden, the Netherlands, the United Kingdom, and the United States. These approaches share a systemic perspective on education. The data that are being collected no longer concern only educational expenditures and enrollments at the national level. A multilevel perspective is applied, and outcome measures, such as graduation rates and measures of student performance in key curriculum areas, are reported in a multilevel framework extending from the individual student to classrooms, schools, localities, regions or states and, eventually, the nation.

The Prospects for the 1990s and Beyond

From the 19th century with legislation in many countries on universal primary schooling, education has been a key element in nation-state building. It continued to have a strong nationalistic orientation throughout the period of postwar reconstruction and development. Today education is still in the service of the nation-state; in some countries even more so than before. However, in Europe and, to some extent, in North America and Asia changes are taking place as a consequence of profound changes in the political geography of the regions. Accordingly, education policy has lost some of its preoccupation with localism and regionalism. These developments naturally have had repercussions also above the national level.

In the late 1980s the Center for Educational Research and Innovation at the OECD undertook a feasibility study aimed at discovering whether it would be possible to develop a limited set of *indicators* of education systems (OECD, 1992a). This politically very delicate exercise was successful in that it eventually, in 1991, led to a decision to produce a limited set of international comparisons on key aspects of the education system. The first international report, *Education at a Glance: OECD indicators* included data concerning the 1988 school year (OECD, 1992b). This volume, as well as its successor, *Education at a Glance II* (OECD, 1993), which included 38 international indicators pertaining to the 1991 school year, became instant bestsellers. Another example of change in the international policy environment of education is the mandate of EUROSTAT, the Statistical Office of the European Union to collect community-wide data on educational inputs, processes and outcomes, and to present this information in a comparative framework based on a set of education indicators. A third example is IEA's Third International Mathematics and Science Study (TIMSS) with participation of some 45 countries. The fact that policymakers have increasingly become aware of the importance of education for the competitive edge of their countries has lent particular importance to the three initiatives mentioned.

These examples show that two important preconditions for the start of a new phase in comparative education are now increasingly satisfied: political support and scientific feasibility. First, recognition of the international dimension in educational policy and evaluation has created a demand for information at the international level. Today there is a distinct political will in several countries to build up a comparative knowledge base in education. Such support was until recently often lacking. Second, there has been a large methodological advance since the first international study of mathematics was conducted by IEA in the early 1960s—examples are new statistical methods for the measurement of theoretical, not directly observable variables, the scaling of achievement scores, and the specification of multilevel models. Building on the work conducted during the past few decades, the field is now approaching the time when a hierarchical, multilevel information system in education can be established. Thus the 1990s bring new and exciting prospects for the development of comparative education.

Why and How Monitoring Came About

The above overview has shown that the history of, at least empirical, educational assessment and the monitoring of national systems of education is rather short and goes back only to the early 1960s. Some answers to the questions raised at the beginning of this chapter can be derived, at least by implication, from the foregoing presentation.

Why Monitor the Standards of Education?

The historical overview suggests that monitoring was called for because it was considered as a necessary part of the knowledge base needed for policy decisions, for example about the allocation of resources. Why, then, be concerned with education standards, and why embark on monitoring exercises? This chapter has provided several possible answers to these questions, some being more important in less developed countries, others in more developed countries. The following ones seem important:

The *costs* for school education in real terms per student and year in the industrialized countries have tended to increase considerably with time whereas outcomes in terms of student achievement have not been enhanced proportionately. This statement can be derived only from time series observations on costs and outcomes. Thus, an important reason for the collection of data to feed a monitoring system is simply to discern important trends in the development of education. Knowledge of major trends is necessary for the framing of education policy, since without such information the school authorities will lack the framework of reference they need for understanding what is happening in the system in terms of inputs, processes and concrete outcomes.

This indicates a second major reason why one should be concerned about the monitoring of education. Systematic *data collection* and analysis are important preconditions for growth in the tree of knowledge. The advance of the educational sciences and especially comparative education has long been hampered by a lack of systematic, computer-based information. The building up of monitoring systems and the aggregation of data at the international level will eventually yield a comparative information base that can be used for the testing of hypotheses about stability and change in education.

From time to time curricular reforms, not to mention *structural reforms*, take place and one needs to know how and to what extent such changes have affected the outcomes of the system. Have standards improved or deteriorated? Again, judging the success or failure of an educational reform presupposes that information on both pre- and post-treatment variables be systematically collected and analyzed.

In countries with a national curriculum and goals for education set by central authorities there is a need for information on how well the education system measures up to the goals. One could call this the *leitmotiv*, the quest for accountability. Its main tenet is that national authorities, school district authorities, or local education authorities, school principals and teachers, are answerable to the public and the users of educational services for the results that are achieved.

In countries with decentralized systems but with certain stated goals and financial support on the part of the central government, the latter is interested in knowing to what extent local schools live up to the goals and the degree of equality of education that is offered. This presupposes that a systematic effort be made to measure the educational process, assess the achieved outcomes and evaluate the findings, so that success and failure can be discovered and, if required, appropriate action to improve the situation be taken. Given the absence in many countries of concrete evidence on the functioning and performance of the system, the implication is that governments must develop and maintain a capacity for the systematic evaluation and monitoring of educational processes and outcomes.

Goals and standards are set because they are believed to carry incentives that may strengthen the achievement motivation of students and teachers. However, for this to work it is necessary that the standards have a real and tangible meaning. If standards are set very high, then some students may not be able to reach them. If they are set too low then they will not have the desired effects in terms of improved incentives and enhanced motivation. If the standards are set at a high level then the weak students will need more time and extra support. Performance assessment and monitoring will be required in all these cases, especially since rewards and sanctions are normally associated with the standards. The sanctions associated with poor performance can lead to tensions between those who set the standards and those who are judged by them.

This leads to the issue of how the organization and processes of schooling are controlled. In many countries—and not only federal ones—various regional or local authorities play a significant role in the determination of education policy. The setting up of a monitoring system for the evaluation of education policy can be seen as an external intervention to direct the efforts of local authorities and steer the internal operations of schools towards objectives deemed important at a national level. A monitoring system has therefore implications for the policy system itself, since it may well serve to strengthen bureaucratic and administrative control over the system. The former Soviet Union wås a case in point.

In summary, the reasons why monitoring came about are closely linked to the perceived functions of monitoring national systems of education. These functions can take several forms, depending on the agents and their purposes:

(1) *Accountability:* enriching public discussion by reporting on the overall status and the strengths and weaknesses of education, thus encouraging the setting of education goals and performance standards.

(2) *Enlightenment*: improving understanding of the functioning of education, and increasing sensitivity to the similarities and differences in the education systems of countries.

(3) *Decision-making:* informing and improving the administration and management of education by identifying system weaknesses, such as the misallocation of resources, facilities and time, personnel inefficiency, and inadequate student performance; by monitoring changes over time in key variables; by providing input into reform planning and implementation; and by assessing the impact of reform efforts.

(4) *Advancement of science:* supporting the development of theories and appropriate methods for the measurement of outcomes in multiple dimensions.

(5) *Administrative control*: influencing the structures, means and ends of decision-making in the education system.

Who Determines Standards, and Who Are the Customers?

The above overview has shown that the process of standard setting is closely linked to that of goal specification, and that the goals for education are framed through historical and cultural processes. Yet the example of the United States shows that goals and standards can also be set deliberately, through a process of political codetermination initiated at a specific point in time. Whereas the goals for education are mostly determined at a political level, their translation into concrete standards and procedures involves a scientific as much as a political process. Standards are informed by public and professional discourse, by the judgments of influential actors, and by international benchmarks derived from international studies. Accordingly, policy considerations, scientific and technical considerations, and practical issues jointly influence the determination of standards. Several criteria thus apply for choosing, developing and evaluating education standards for use in a monitoring framework (adapted from Nuttall, 1993):

(1) Standards are anchored in goals and objectives;
(2) Standards are anchored in central features of the education system, and they have a bearing on current or emerging policy issues and problems;
(3) Standards are both diagnostic and judgmental;
(4) Standards must be accessible to all stakeholders;
(5) Standards must be quantifiable and imply the measurement of observed behavior rather than perceptions;
(6) Standards must be valid and reliable;
(7) Standards should have general applicability;
(8) Standards should be useful in comparisons;
(9) Standards should be implementable.

The overview has given some insights into who the "customers" are of a monitoring system in education. In the first place they are politicians and decision-makers in parliaments and governments, secondly administrators and educational planners in central agencies. Typically, educational planning as a systematic exercise was almost unheard of until about 1960 when OECD was established under its present name. At about the same time the World Bank and the Ford Foundation in conjunction with UNESCO supported the founding of the International Institute of Educational Planning in Paris. Thirdly, public opinion is served by information emanating from monitoring. A fourth user category is made up of the practitioners in school systems, mainly school principals and teachers. Improved information about the factors that influence student performance and the processes of schooling—in so far this understanding leads to improved decision-making and improved practice—is also assumed to benefit the students themselves. Finally, the research community is an important customer group. A monitoring system requires that data be collected. Whereas governments can partly satisfy the need with data emerging from administrative records, increasingly survey data are also called for. Here researchers have a role to play. Moreover, researchers have an interest in data analysis, and the resulting advance in knowledge and understanding may be assumed to benefit the field. Finally, a monitoring system is not static; it needs continued research for improvement.

The problem of how standards can be set and monitoring conducted is addressed in subsequent chapters. Whereas decisions about the scope and costs of a monitoring system are generally taken at the political level, issues of implementation, operation and even use are mainly decided by decision-makers and administrators in government agencies. Researchers play a limited but crucial role; they address the questions about how monitoring technically can be conducted. But the way monitoring is implemented and the uses that are made of the findings is steered by the intentions of policymakers and the concrete issues they face. These issues vary from country to country, something that can be studied when looking at the way the outcomes of the IEA surveys were utilized in different countries (Husén, 1979; Husén & Kogan, 1984; Husén, 1986).

What are the Implications?

Monitoring has pervasive implications for school systems, some of which may be judged as positive and others as negative. Whereas the functions of monitoring seem straightforward, implementation is not. If properly used, a monitoring system may help achieve the implied purposes. But the potential for misuse is high, too. Being responsible for the invention and operation of a monitoring system thus by necessity brings a certain responsibility for limiting the scope of misuse. Below a few possible ways of misuse are mentioned.

First, there is the risk that a monitoring framework may be used for reasons of instrumental control. Decision-makers might use the information to legitimize their holding control over schools and teachers. This, of course, is not the intention, since the primary purpose of the new information is to foster an informed, sustained and also critical discourse about the means and ends of education. The risk that a monitoring system be misused in this way can be reduced if an independent reporting strategy is defined that focuses on the accountability of political administrations *vis-à-vis* their constituencies.

A second point is that monitoring systems usually are developed through a process of consensus seeking. This implies that authority rests on a set of shared views. Monitoring moreover presupposes goal and standard setting, and this in turn implies a choice as to what issues are considered more important than others. Goals and standards thus focus attention on some specific outcome measures—often indicators of student achievement—while the other goals and objectives of schooling may be given less attention. This raises the problem of determining which outcomes to measure. For pragmatic reasons, it is tempting to limit a monitoring system to the data that are available or that can be collected at a reasonable cost. Clearly, if a monitoring system were to rely solely on data collected through standardized multiple-choice tests in a few major subjects, little advance in understanding would be achieved. Worse, such a situation might focus attention in undesirable ways and drive curriculum development in a misguided direction (Bryk & Hermanson, 1994). But schools seek to instill values and foster outcomes in addition to those that are easily quantifiable. The challenge, therefore, is to develop a more balanced notion of educational outcomes. Another implication is that a favorable research policy should accompany the development of a monitoring system.

Attempts to monitor a given national system and to measure its "educational productivity" would also have to take into account the socio-economic and ethnic changes that occur over relatively short periods. The report on schooling in modern European society by a task force set up by the *Academia Europaea* (Husén, Tuijnman & Halls, 1992) highlights, for instance, the transformation that the family has undergone in recent decades as well as the demographic changes caused by the immigration of labor. In some European countries the number of school children coming from homes where the national language is not spoken has multiplied since the early 1960s.

Conclusion

Assessment and monitoring depend on measurement techniques, and the findings derive their meaning from a comparative context. Limitations are thus imposed, because the state of measurement technology is arguably always imperfect, and also because theory has not advanced to the point where contextual information of a more qualitative nature can be fully build into the data analysis. However, monitoring a national system of education in an international setting by means of empirical methods is not a meaningless exercise. Taking socio-cultural differences into account, international comparisons can be extremely illuminating. But in so doing, survey research methods have to be complemented by other, qualitatively oriented techniques, for example as was done by Stephenson and Stigler (1992) in their studies of classrooms in the United States, Japan, and China. This study shows that great care is needed in interpreting student assessment data, because of differences in the historical, cultural and social frame factors of the examined school systems. Discourse, openness and collaboration among social science disciplines and research teams are needed in order to avoid conceptual distortion.

The inability to accommodate dissent and argument was, in fact, a powerful reason for the demise of the social indicators movement in the early 1970s. The technical problems as identified by experts in educational testing and measurement should not be allowed to overshadow the political, economic and sociological aspects that must be considered if monitoring results are going to be fruitfully interpreted. Failure to do so might well trigger a new crisis of confidence in the value of educational assessment and monitoring. To avoid this, ways of uniting both the research field and the producers and consumers of research evidence must be found (Tuijnman, 1993).

References

Aikin, W.M. (1942). *The story of the eight-year study, with conclusions and recommendations.* New York: Harper and Row.

Bauer, R. (Ed.) (1966). *Social indicators.* Cambridge, Massachusetts: MIT Press.

Beeby, C.E. (1977). The meaning of evaluation. In *Current Issues in Education, No. 4: Evaluation* (pp. 68-78). Wellington: Department of Education.

Bestor, A. (1953). *Educational wastelands.* New York: A. Knopf.

Bloom, B.S., Madaus, G.F., & Hastings, J.T. (1970; 1981). *Evaluation to improve learning.* New York: McGraw-Hill.

Bryk, A., & Hermanson, K. (1994). Education indicator systems: observations on their structure, interpretation and use. In A.C. Tuijnman & N. Bottani (Eds.), *Making education count: Developing and using international indicators.* Paris: Paris: OECD, Center for Educational Research and Innovation.

Coleman, J.S., *et al.* (1966). *Equality of educational opportunity.* Washington, DC: US Department of Health, Education, and Welfare; Office of Education.

Foshay, W.A. (Ed.) (1962). *Educational achievements of 13-year-olds in twelve countries.* Hamburg: UNESCO Institute of Education.

Husén, T. (Ed.) (1967). *International study of achievement in mathematics: A comparison of twelve countries. I-II.* Stockholm: Almqvist & Wiksell; New York: John Wiley.

Husén, T. (1979). An international research venture in retrospect: The IEA surveys. *Comparative Education Review* 23(4), 371-385.

Husén, T. (1986). Moderne Psychologie und das deutsche Bildungswesen: Rückblick auf eine internationale Arbeitstagung 1952 in Frankfurt am Main. *Zeitschrift für internationale Erziehung und sozialwissenschaftliche Forschung* 3(1), 17-30.

Husén, T., & Kogan, M. (Eds.) (1984). *Educational research and policy: How do they relate?* Oxford: Pergamon Press.

Husén, T., Tuijnman, A.C., & Halls, W. (Eds.) (1992). *Schooling in modern European society: A report of the Academia Europaea.* Oxford: Pergamon Press.

IEA (International Association for the Evaluation of Educational Achievement) (1988). *Science achievement in seventeen countries: A preliminary report.* Oxford: Pergamon Press.

Keeves, J.P. (Ed.) (1992). *The IEA study of science III: Changes in science education and achievement: 1970 to 1984.* Oxford: Pergamon Press.

Livingston, I.D. (1985). Standards, national: Monitoring. In T. Husén & T.N. Postlethwaite (Eds.), *International Encyclopedia of Education* (pp. 4786-4791). Oxford: Pergamon Press.

Nuttall, D.L. (1994). Choosing indicators. In A.C. Tuijnman & N. Bottani (Eds.), *Making education count: Developing and using international indicators.* Paris: OECD, Center for Educational Research and Innovation.

OECD (1973). *A framework for educational indicators to guide government decisions.* Document for general distribution. Paris: OECD.

OECD (1992a). *The OECD international education indicators: A framework for analysis.* Paris: OECD, Center for Educational Research and Innovation.

OECD (1992b). *Education at a glance: OECD indicators.* Paris: OECD, Center for Educational Research and Innovation.

OECD (1993). *Education at a glance: OECD indicators. 2nd edition.* Paris: OECD, Center for Educational Research and Innovation.

Postlethwaite, T.N., & Wiley, D.E. (1992). *The IEA study of science II: Science achievement in twenty-three countries.* Oxford: Pergamon Press.

Robitaille, D.F., & Garden, R.A. (1989). *The IEA study of mathematics II: Contexts and outcomes of school mathematics.* Oxford: Pergamon Press.

Schultz, T.W. (1961). Investment in human capital. *American Economic Review 51,* 1-17.

Stephenson, H.W., & Stigler, J.W. (1992). *The learning gap: Why our schools are failing and what we can learn from Japanese and Chinese education.* New York: Summit Books.

Super, D.E. (Ed.) (1969). *Toward a cross-national model of educational achievement in a national economy.* New York: Columbia University and Teachers College Press.

Svensson, N.-E. (1962). *Ability grouping and scholastic achievement: Report on a five-year follow-up study in Stockholm.* Stockholm: Almqvist & Wiksell.

Tuijnman, A.C. (1993). Themes and questions for a study on educational research and development. In T.M. Tomlinson & A.C. Tuijnman (Eds.), *Educational research at the crossroads.* Washington, DC: Office for Educational Research and Improvement, US Department of Education.

Tyler, R.W. (1989). *Classic works of Ralph W. Tyler,* compiled and edited by G.F. Madeus and D.L. Stufflebeam. London: Kluwer Academic Publishers.

US, Department of Education (1991). *America 2000: An education strategy.* Washington, DC: US Department of Education.

US, Department of Education (1993). *Education in the states and nations: Indicators comparing US states with the OECD countries in 1988.* Washington, DC: National Center for Education Statistics, US Department of Education.

US, National Academy of Education (1993). *Setting performance standards for student achievement.* Stanford, CA: The National Academy of Education, Stanford University.

US, National Commission of Excellence in Education (1983). *A nation at risk: The imperative for educational reform.* Washington, DC: US Department of Education.

US, National Council on Education Standards and Testing (1992). *Raising standards for American education: A report to Congress, the Secretary of Education, the National Education Goals Panel, and the American people.* Washington, DC: US Government Printing Office.

US, National Education Goals Panel (1991). *Building a nation of learners: The national education goals report, 1991.* Washington, DC: National Education Goals Panel.

US, National Education Goals Panel (1992). *The national education goals report, 1992.* Washington, DC: National Education Goals Panel.

US, Nationel Education Goals Panel (1993). *The national education goals report. Volume one: The national report, 1993.* Washington, DC: National Education Goals Panel.

US, The White House (1993). Goals 2000: Educate America Act. Legislation submitted to the US Congress by William J. Clinton, April 21, 1993. Washington, DC: Office of the Press Secretary.

Walker, D.A. (1976). *The IEA six subject survey: An empirical study of education in twenty-one countries.* Stockholm: Almqvist & Wiksell, and New York: John Wiley & Sons.

Wyatt, T. (1994). Development of education indicators: A review of the literature. In A.C. Tuijnman & N. Bottani (Eds.), *Making education count: Developing and using international indicators.* Paris: OECD, Center for Educational Research and Innovation.

Chapter 2

Monitoring and Evaluation in Different Education Systems

T. NEVILLE POSTLETHWAITE

Institute of Comparative Education, University of Hamburg, Germany

Different education systems employ different monitoring and evaluation practices. This chapter first describes the kinds of inputs, processes, and outcomes of education that are monitored by many ministries interested in evaluating the achieved results in terms of desired standards. Second, examples are given of how ministries monitor the level of inputs, processes and outcomes in schools in terms of specified standards being reached. Third, a brief description is presented of the way in which ministries of education organize their data collections. Finally, a few suggestions are made on how certain aspects of international studies of educational achievement might be improved if the comparability of national indicators is to be ameliorated.

This chapter describes the regular and systematic monitoring and evaluation of inputs, processes, and outcomes in school systems as undertaken by both national and international agencies of education. It does not address monitoring that is conducted to evaluate specific reforms or changes that have been introduced into education systems. As used in this chapter, monitoring refers to the regular collection of information about different aspects of the education system, and evaluation means the placing of a value or judgment on the data collected. Much of what has been written in Chapter 1, and indeed in some of the subsequent chapters, has focused on international monitoring and on evaluation studies with an emphasis on outcome indicators. Hence, this chapter will focus more on what national ministries of education undertake in collecting data about inputs into schools as well as about processes and outcomes. First, what is typically monitored will be described. Second, the issues of levels and equity of the inputs, processes and outcomes will be examined. Third, examples of how countries organize their data collections will be presented and, finally, some comments will be made about selected aspects of international monitoring and evaluation exercises.

What is Monitored?

It is possible to classify what is monitored into three categories: inputs into schools, processes in school, and outcomes of schooling. Table 2.1 presents a few examples of the kinds of indicators or variables in which ministries of education are often interested. Different ministries monitor different aspects of their school systems. Not all of the aspects of an education system mentioned in Table 2.1 are of interest to all education systems. It is suggested that those aspects of a system that are believed to be "in order" are not monitored. That is, the system as now constituted is perceived to work adequately and hence there cannot be any serious problems. On the other hand, where the public, political factions, or administrators of the system believe that there are components that are causing serious problems, then the monitoring of those aspects can be undertaken. Thus, what will be monitored can differ somewhat from one system to another, depending on factors such as the maturity of the system, the system's financial condition, the quality of teacher preparation, the public's satisfaction with the achieved output, and the perception of the system's administrators of where the problems might lie.

In some cases, it is debatable whether or not an indicator should be classified under inputs, processes, or outcomes. This is a matter of perception, and readers' perceptions will naturally differ. For example, should absenteeism be classified as a process or an outcome variable? Or, is class size an aspect of an education input or process? The rest of this section describes, in more detail, some of the input, process and outcome indicators commonly considered in national monitoring systems.

Inputs into Schools

It is the responsibility of education authorities to provide school buildings, maintain them, equip them and ensure that they receive school supplies so that they can operate. The authorities are also responsible for supplying the schools with teachers and, at some levels, administrators. In some countries, social and health services and functions such as the provision of school meals are considered to be societal inputs into the school system.

There is sometimes a difference between those who want the information about school inputs and those who are responsible for their provision. For example, in Hungary, the local municipalities are responsible for the school buildings and their equipment. Even though the ministry of education has regional offices of education, they are responsible only for the pedagogical aspects of what transpires within the schools under their jurisdiction. Since, in 1993, the ministry was concerned about the level of inputs it took it upon itself to collect the information and then inform the local municipalities if and where action was required. The same was true in Zimbabwe for a number of years but the responsibility for all buildings, equipment, and supplies has now reverted to the regional offices of education.

Table 2.1. Examples of indicators of interest to ministries of education

Type of indicator	Examples of indicators
Inputs	Condition of school building
	Condition of teacher housing
	School furniture
	School supplies
	School laboratories
	Total number of pupils
	Age, grade and sex of pupils
	Number of full-time or equivalent teachers
	Pupil:teacher ratio
	Class size
Process	Teacher work load (hours of instruction per week)
	Teacher perceptions of factors influencing instruction
	Curriculum (nationally, regionally, school prescribed)
	Opportunity to learn
	Hours instruction per subject per grade level
	Number of pupils studying which subjects per grade
	Inspectors' visits (how many per term)
Outcomes	Achievement in key subjects at major points in system
	Percentage of grade group graduating
	Percentage students obtaining examination results
	Expectations and attitudes of pupils
	Absenteeism
	Violence
	Drug use
	Discipline problems

School buildings. What type of data are required about school buildings? These differ from country to country. For example, in Zimbabwe, it was and still is important to know how many classrooms there are in each school and how many more are still required; furthermore, the ministry is interested to know how many classrooms have a roof and no walls (or walls and no roof) and the materials with which the classrooms are built (mud, wood, stone, brick, etc.), whether there is electricity, and whether there is running water, water from a bore hole or no water at all. In some countries, the number of toilets available for boys and, separately for girls is a problem. In these cases it is the number of boys and girls per toilet that is required for each school. The reason why Zimbabwe wanted these data was because the ministry had minimum specifications of what a school should possess and therefore wanted to know to what extent these specifications were being met. These specifications (or standards) are known as norms in Zimbabwe. In other school systems, information is required about the number of broken windows each year; in certain cases, the reasons for so many broken windows are also sought.

Equipment. What data are collected about equipment? Again, this differs from system to system. To take Zimbabwe again, the norms require that every classroom should have one blackboard, one storage cupboard, one meter of bookshelves, writing places and sitting places for all pupils in the class—there was meant to be a maximum of 40 pupils per classroom and hence one object of the data collection exercise was to establish how many pupils there were in each classroom. In other countries, the information required deals with the existence of a classroom library in primary schools, the number of books in the library, and the extent to which pupils can borrow the books to take home and read. In secondary schools, similar information is sought about the school library and, in some cases, about the number of new titles added to the library in a given year.

The existence of different types of science laboratories and sports facilities, and how often they are used for practical as well as theoretical work, are additional examples of data that are frequently collected.

Supplies. Many ministries have norms for supplies. For example, each pupil should receive one notebook per term, or each pupil should have one textbook for each subject being taught, or so many ball-point pens per term. Textbooks are perceived as being very important since there is a severe lack of them in many countries. Research findings indicate that, for example, where two rather than five pupils share a textbook, their level of achievement rises. Given that a ministry has these minimum standards, it wishes to assess to what extent these standards are being met.

Teachers. Ministries require different sorts of information about teachers. These include the number of full-time teachers and full-time-equivalent teachers (for example, two half-time teachers equal one full-time teacher). This allows, for example the calculation of the pupil:teacher ratio in each school. The average of these figures for each school result in the average pupil:teacher ratio for the country or for different types of schools. Information on the sex of the teacher is also often collected. In some countries, there are regulations stating that there should be a 50/50 balance between male and female teachers. Ministries need to know to what extent the regulations are being followed. Information is also collected on the age of each teacher. This datum is collected only to be able to calculate the age distribution of teachers within a school (quite often to ensure that the teachers within a school are not all too old or too young). In countries where there have been many different teacher training programs and hence different types of qualifications with some being regarded as superior to others, a ministry often wants to know the distribution of these qualifications within each school. In systems of education where the ministry allocates teachers to schools this is particularly important. However, in systems where the school appoints teachers and the ministry has no say in the matter, such data are mostly not collected.

Decision-making. A popular indicator is that of the locus of decision-making. It is becoming recognized that neither top-down orders nor bottom-up decision-making are desirable, but that different levels in the administrative hierarchy need to be conjointly involved in decisions about many aspects of schooling. To this end, some systems are beginning to collect data on the different levels in the school system that are concerned with such areas as resource allocation and use, curriculum decision-making, and the pedagogical organization of pupils within the schools. These data are usually collected in terms of the extent to which teachers are involved in decision-making. However, it must be said that the collection of this type of information tends to be the subject of special rather than regular data collections. Finally, some regard teacher salaries as a conditioning factor for the attraction of good students to the teaching profession, and in international studies much attention is paid to the relationship between teacher salaries and student achievement.

Pupils. All ministries collect data about the number of pupils (boys and girls separately) in the school and their ages. These data are typically collected by grade and often by class. This allows the computation of the age by grade distribution within each school, pupil:teacher ratio, class size, and school size. A few ministries do not collect the age data but these are exceptions to the rule. It may be expected that all ministries will soon be collecting these data.

It should be noted, however, that ministries are beginning to link these input factors to outcomes in terms of what pupils learn. The aim is to identify those input factors that have an impact on student achievement, so that efforts can be focused on dealing with those aspects first.

Processes in Schools

Teacher time. Teacher work load or the number of hours each teacher stands before the class is often considered to be an important piece of information. Similarly, information on the number of hours each teacher reports spending preparing lessons and correcting homework is collected, even though the problem of the validity of self-reports for this type of information is well known. In some countries, ministries are interested in the views of teachers about their working conditions, their living conditions (where the ministry is responsible for teacher housing, as is the case in many developing countries), and the distance (or time needed) to commute from their homes to the school each day. Such data are then used by the ministry to reallocate existing resources or acquire new resources to ameliorate the living conditions or other aspects of the teachers' life. In some countries, an effort is made to collect data about the views or perceptions of teachers about the conditions in the school and which of these they believe affect their teaching. Where there is a high degree of agreement among the teachers, the ministries then attempt to improve those particular conditions.

Curriculum. Curriculum can be viewed as an input or as a process. Here it is classified as a process but where there is a national curriculum, the curriculum could also be viewed as an input into the schools. In many countries the national curriculum is viewed as a minimum coverage and the schools or teachers can add more. Or, in some countries, there is a national part which is meant to cover 60 or 80 percent of the time allocated to a particular subject matter, and the other 40 or 20 percent is meant to be for regional or even school-determined purposes. Ministries are often interested in assessing to what extent the national curriculum is being covered by the teachers and hence they collect information on this. Indeed, one of the major tasks of inspectors in some countries is to check on the curriculum coverage and opportunity to learn.

Opportunity to learn. There is a growing feeling that "opportunity to learn" (OTL) is an important "cause" of learning. Bluntly stated, the concept is that if pupils are offered the opportunity to learn something they will, in general, learn it; if they are not offered the opportunity to learn, they will not be able to learn it. Thus, there is a growing effort to collect valid information for this concept in each class in each school. It does not require much perspicacity to see that this is not an easy concept to measure. Pupils change schools and teachers within schools. Who should furnish such information—the pupil or the teacher, but which teacher? The International Association for the Evaluation of Educational Achievement (IEA) first attempted to measure OTL (cf. Husén, 1967). The IEA measure accounted for differences in achievement between countries but was much less successful in accounting for differences in performance between pupils within a country. Development work has been proceeding in the United States (cf. Schmidt, 1993a and 1993b), but it remains to be seen to what extent this can and will be used by ministries of education in their routine data collections.

Time allocation. In some countries there is a central timetable that is laid down for each grade level; it is often a decree of Parliament. In other systems, the school timetable is decided by each school. In this latter case, ministries often wish to have information on what is happening in the school. Therefore, they collect information on the number of hours and minutes per week and school year that are devoted (at least according to the timetable) to each subject at each grade level.

School organization. There are the issues of how the school organizes its classes (homogeneous or heterogeneous ability grouping), the amount of grade repeating that is used, the assessment procedures used, and the school's and the teachers' contact with the parents and with the community in general. Some ministries collect data on these matters but most do not. Even when data are collected, the data collection exercise tends to be a special one and not part of the yearly data collection.

Subjects offered and pupils studying them. At the secondary school level, each school covers the stipulated national curriculum, if applicable, but the extra curriculum offered within each school is often not known. For example, by the end of compulsory schooling in some European countries there are schools that offer several foreign languages. There is often no knowledge at the national or regional levels about how many pupils are studying which foreign languages, and how choice of foreign languages is changing over time. This information is required to plan the supply of teachers in the different subjects. It should also be pointed out that without such information it is extremely difficult to draw national probability samples of those studying any or each of these languages for monitoring purposes.

Inspection. In certain countries, it is expected that the local inspectors will visit each school within their jurisdiction at least once a year and carry out their tasks. Apocryphal stories abound about the inspectors who go only to visit the school principal for a cup of tea or coffee and never enter the classrooms or speak with the teachers; and, of schools that have never seen an inspector for five years. An effort is, therefore, made by ministries to collect data about inspectors' visits from sources other than the inspectors themselves.

The extent to which ministries gather data on several or all of these issues seems to depend on the amount of public concern about them. However, this is not the case with methods of teaching which are often assessed through school inspections.

Outcomes and Outputs

Outcomes are normally concerned with pupils' cognitive achievement at selected age or grade levels in the school system. Cognitive achievement is broadly interpreted and includes all levels of the Bloom taxonomy (Bloom, 1956), or the SOLO taxonomy (Biggs & Collis, 1982), extending from factual memorization to the solution of complex problems. However, other skills such as communication, cooperation, and the ability to work with others are sometimes also assessed. Occasionally selected attitudes and values are measured. Some systems at times conduct work to assess pupils' attitudes such as "like-dislike school", "like-dislike particular subjects", "perceptions of the quality of school life", and "democratic values". The reader is referred to the chapter by Judith Torney-Purta in this book for insights into the monitoring of attitudes and perceptions.

Achievement is usually assessed at the end of primary school, the end of compulsory schooling, and the end of secondary schooling. The outcomes measured are usually in the core subjects. At the primary school level, the subjects are the core or key subjects and include mother tongue, mathematics (usually arithmetic), and sometimes social studies and science. At the secondary school level, they include mathematics, science (biology, chemistry, physics, and sometimes earth sciences), mother tongue, and one or more foreign languages. In some systems, history and geography are also included.

The aim of the monitoring and evaluation of outcomes appears to be twofold: first, to identify those aspects of each subject matter that are being well achieved and poorly achieved, and second, to identify if achievement levels are remaining constant over time or are improving or deteriorating. In particular, those responsible for the system are interested in the particular aspects of a subject matter where achievement is deteriorating so that, if they are important aspects, action can be taken through curriculum development or through teacher training to improve them.

Outputs are the number of pupils reaching a selected level of schooling or, in other words, the number of persons progressing through the school system. What percentage of an age group actually reach particular levels of the school system? What amount of grade repeating takes place? This number is expressed either as a percentage of an age group or as a percentage of the number of pupils that entered first grade. It is, of course, one thing to report that 84 percent of an age cohort reaches the final grade of schooling in one system and 40 percent in another. It is another thing to report that the first group has, say, an achievement "quotient" of 70 and the second a "quotient" of 75. In other words, it is important to know how many of the pupils know how much of the content to be learned. This is what the quotient (sometimes called 'yield') reports. It is clear that a system could bring a lot of pupils to the final grade of school but they may not have learned very much (Postlethwaite, 1967).

In some school systems there is an increased interest in having measures of truancy and the reasons for different types of truancy. Associated with this, there is the problem of the frequency and types of disciplinary problems in the schools. Whether these two problems are best classified under outputs or processes is debatable. Since these problems are typically monitored through special research studies rather than through regular and systematic monitoring and evaluation studies, no further mention will be made of them.

Levels and Equity

Measures of Levels of Inputs

As already mentioned, ministries of education are interested in the level of inputs into schools and whether or not those inputs are the same in all regions of the country and in all schools in all regions. Where the specified level is not reached, then the ministry knows that it must do something. Similarly, if there are disparities, then action is required if the policy is to have equal inputs allocated to all schools. It must be borne in mind that, in some countries, there is a deliberate policy of differential inputs into schools in the sense that deprived schools are often given more resources.

In a study conducted in Zimbabwe in 1991, the Minister of education specifically asked for the level of selected inputs to schools to be examined. Table 2.2 presents the ministry's norms for the items in the list as well as the percentage of pupils in classrooms achieving or exceeding the standard (Ross & Postlethwaite, 1992).

Table 2.2. Percentage of grade 6 classes achieving or exceeding government norms in Zimbabwe, 1992

Furniture and supplies	Ministry norms	Percentage classrooms equal to or above the standard	Standard error of sampling
Blackboard	1	92.9	2.13
Seats	one per pupil	18.4	3.22
Desks	one per pupil	11.5	2.65
Class size	40	37.2	4.01
Cupboard	1	43.2	4.11
Bookshelves	1	35.9	3.98
Exercise books	3	70.3	3.79
Notebooks	1	11.0	2.60
Pencils	1	24.8	3.59
Rulers	1	27.6	3.71
Erasers	1	7.1	2.13
Ballpoint pens	3	0.7	0.22

Source: Ross & Postlethwaite (1992).

The final column in Table 2.2 presents the standard errors of sampling (since the data collection was based on a probability sample of schools, classes, and pupils). It can be seen that the ministry's standard for blackboards was one blackboard per classroom and that 92.9 percent of pupils in Grade 6 classrooms (primary school) were in classrooms that attained the norm. One standard error of sampling for this percentage based on the sample was 2.13, indicating that the sample percentage, with 95 percent confidence, was between 90.8 percent and 95.0 percent. While most pupils were in classrooms that achieved the standard for a blackboard, it can be seen that in general most inputs did not attain the standards set by the ministry.

It is possible to use these data to compute other statistics that can be of use to the ministry. Table 2.3 presents information based on composite indicators for both classroom furniture and classroom supplies. This was done by allocating a score of 1 to each classroom that met the standards and a score of 0 to each classroom that did not. The items were then aggregated. There were five items for furniture (class size was omitted), thus making a five point scale. There were six items of supplies making a six point scale.

For furniture it can be seen that 6.5 percent of classrooms had none of the items, 32.0 percent had only one of the items and so on.

Table 2.3. Percentage frequency distributions for composite indicators of classroom furniture and supplies, Zimbabwe, 1992

Indicator value	Furniture	Standard error of sampling	Supplies	Standard error of sampling
0	6.5	2.05	24.4	3.57
1	32.0	3.87	41.5	4.09
2	33.9	3.93	14.4	2.92
3	13.9	2.87	11.9	2.69
4	8.3	2.29	4.3	1.68
5	5.5	1.89	2.8	1.37
6	0.0	—	0.7	0.69

Note: — not applicable.
Source: Ross & Postlethwaite (1992).

The above two tables present the data for Zimbabwe as a whole. There are nine educational regions in Zimbabwe and it would also be possible to present the data by region. This is done in Table 2.4.

The large number of blanks in the bottom half of Table 2.4 indicate that there is a problem with the standards for supplies being met in nearly all regions of the country.

Measures of Equity of Inputs

One question of interest to those responsible for the operation of education systems is if the variation in provision is more between regions or more between schools within regions. Ministries sometimes use a statistic that is known as the coefficient of intraclass correlation (rho) (Postlethwaite, 1994) to estimate variance between regional mean scores as a component of the total variation between the mean scores of the schools. If the rho is high, then there is large variation between regions and if it is low, the variation is mostly between schools within regions. The rhos for Zimbabwe for the two composite variables of furniture and supplies were 0.35 and 0.06 respectively. This statistic must be interpreted together with the overall levels of provision. In the case of Zimbabwe it was clear that the overall level of provision was low. In general, if the rho is high, as is the case for furniture, then there is substantial variation between regions, indicating that it may be the task of the ministry to equalize the provision among the regions. If the rho is low, then it is for each region to equalize the provision to all schools within the region. However, if the overall provision is low, then it is the responsibility of both the national and regional authorities to review and act on the problem of provisions. In the Zimbabwe case, the decision was taken to have the national level deal primarily with the problem of furniture and both the national and regional levels to deal with supplies.

Table 2.4. Percentage frequency distribution of supplies indicators by
region, Zimbabwe, 1992

Supplies indicator value	Region								
	1	2	3	4	5	6	7	8	9
0	13	32	27	20	20	20	47	29	16
1	73	47	36	40	47	20	27	36	47
2	13	11	18	27	20	—	20	29	11
3	—	11	—	7	—	38	7	7	21
4	—	—	9	—	7	15	—	—	5
5	—	—	—	7	7	10	—	—	—
6	—	—	9	—	—	—	—	—	—

Note: — Not applicable.
Source: Ross & Postlethwaite (1992).

The rho statistic—or sometimes a similar statistic known as a Gini coefficient (Johnstone, 1981)—can be used for examining equity at different levels of an education system. In Zimbabwe, it was between and within regions. In other countries, for example Zambia, it might be between and within regions for one calculation and between and within districts within regions for another (Silanda & Tuijnman, 1989).

The statistic can be used for many variables and this author has seen it used for enrollment data, teacher perceptions of school problems (where data have been collected from teachers), the existence of shift systems, teacher qualifications, and school building conditions.

Measures of School Processes

The information regularly collected about school processes is usually carried out by the inspectorate and occasionally by means of ad hoc sample surveys. There is, however, a trend for the school census units to collect yearly information on the timetable for each grade within each school, and to analyze the average amounts of time allocated to the different key subjects. Schools that are very deviant from the average are then visited to discover the reasons for their being very different and, if necessary, corrective action is taken. The number of pupils at the secondary school level studying different subjects is also being introduced into the regular data collections. These trends are noticeable in those systems of education where schools are connected by computer to the regional or national offices of education. Where systems still depend on questionnaires to be completed by schools these more detailed data are still not collected. There is also a movement to collect data on the extent to which schools offer their students the opportunity to learn the different aspects of key subject areas. By 1993, such data had been collected in ad hoc surveys. To what extent measures of opportunity to learn will be included in regular achievement surveys remains to be seen.

Much of the work concerning processes within schools is undertaken by school inspectors and district education officers. Education systems tend to have two levels of inspectors: national and regional or local. In some systems it is the District Education Officer who also acts as the local inspector. An attempt is made to have a full inspection of every school once every five years although in some countries this may be only every ten years. This is carried out by the national inspectorate and involves an inspection of the teaching of every subject matter and every teacher. A written report is then made of the inspectors' comments and this report goes both to the school principal, the governing board of the school (where there is one), to the regional director of education, and to the Minister of education. Although the inspectorate can suggest to the Minister that the school should be closed "because it is not up to the ministry's desired standard", or that a teacher should no longer be employed, by the beginning of the 1990s it increasingly seemed to be the case that the inspector's role had become much more one of giving advice to the teachers and school principals on how they might improve the education being offered in the school. The full inspections tend to be thorough and include comments on such matters as the organization of the teaching of each subject matter, the extra activities conducted by the school, the extent to which the "official" curriculum is being covered, and on matters such as discipline and truancy or drop out in the school. In some cases, the inspectors will collect data by means of a checklist on the conditions of the buildings, furniture, and supplies. These data are then entered onto files. The problem with this approach is that it is sometimes slow, expensive and inaccurate. This is why ministries have moved more to having such data collected by means of sample surveys conducted by the census bureau or by other units outside the ministry.

The local inspectors in many countries visit the schools in their areas more frequently than every five years. Even in large areas an attempt is made to visit each school about once every one to two years. Their role is to inspect individual teachers and to make suggestions to them about improvements in their teaching techniques.

The data emanating from the inspectors' work is rarely in a quantified form and this makes it impossible to conduct the types of data analyses about level and variation presented under the discussion of inputs above.

Ad hoc surveys are sometimes conducted on the processes within schools. Such surveys involve a great deal of observation work. Observation schemas are developed to allow the observers to record not only the organization of the school and the teaching of different subject matters but also exactly what teachers "do" in terms of the structuring of the teaching of a lesson but also the interaction between the teacher and the pupils and the interaction of pupils with other pupils, the clarity of exposition of the teachers, and many other facets of the teaching process. Levels and variations are calculated but without a stated standard about such matters from the ministry it is impossible to calculate to what extent standards are being met or not being met. Some of the surveys also measure pupils' achievement in specific subject areas and correlations are calculated between the process variables and the pupils' achievement. Those responsible for the studies then make statements based on correlational analyses about certain process variables being more related than others to achievement.

The "effective schools" movement grew rapidly in the 1980s (cf. Bosker, Creemers & Scheerens, in press). Given that the home backgrounds of pupils are usually related to the achievement of children, and given that schools serve neighborhoods that vary in the levels and variation of home backgrounds, it is argued that it is easy for schools with pupils from good home backgrounds to have high achievement but that this is much more difficult for schools having children primarily from poor home backgrounds. Thus, based on the relationship between home background for all pupils in the sample and their achievement, it is possible to predict what level of achievement a school should have obtained (see Postlethwaite & Ross, 1992 for an example of how this is done). Those schools that have a mean score above what they should have obtained are called "more effective schools" and those obtaining a mean score below what they should have obtained are called "less effective schools". The difference between the actual scores and the predicted scores are known as the residual scores. In some cases the relationships between the variables measured in the study and the residual scores are calculated and as a result of the data analyses, statements are made about those variables associated with "effectiveness". In other cases the "most effective" and the "least effective" schools are compared, and again statements are made about the variables or indicators that are most different between the two groups of schools. It often occurs that the effective schools are then studied in depth, using qualitative research approaches, to identify other factors that may be predicting the higher than expected level of achievement.

Outcome Measures

Which score constitutes a standard? Two approaches are taken in the monitoring of standards. The first is where the ministry or some body of expert judges sets a standard before the data are collected in what might be called the a priori approach. The second is the derivation of standards after the data have been collected in what might be called the a posteriori approach. One consistent finding in the research literature is that the different standard-setting methods produce different results (cf. National Academy of Education, 1993).

The a *priori* approach is where a standard is set down by the ministry (or other body) before the data are collected. Again, Zimbabwe will be used as an example (Ross & Postlethwaite, 1992). A reading test had been developed for Grade 6 pupils based on the curriculum. It was trialled and several revisions were made before the final version was ready. At this point the ministry of education formed two groups. The first was composed of the ministry's reading specialists. The second was composed of practicing grade 6 teachers. Each group selected those items from the test which pupils would absolutely need to answer correctly to be able to profit from instruction in grade 7 (the last grade of primary school). Out of a test of 59 items, 33 items were selected. Both groups settled on two standards or levels. The first, which was a minimum standard, was that the pupils should be able to master 50 percent of the 33 items. The second, which was a desirable standard, was that the pupils should be able to master 70 percent of the 33 items. For Zimbabwe as a whole 38.1 percent of all pupils reached the minimum standard and 13.8 percent reached the desirable standard. In other words, 61.9 percent did not reach the minimum standard and 86.3 percent did not reach the desirable standard. It is clear that the Zimbabwean students performed much lower than expected by the expert groups.

The *a posteriori* approach is to take the whole test, examine the distribution of scores and, by judgment and empirical approaches, determine the cut-off points on the distribution which, for example, best differentiate the "good" from the "average" from the "poor" readers. Had this been done with the Zimbabwe grade 6 reading data, then it would have turned out that 20 percent were good readers, 50 percent were average readers, and 30 percent poor readers. This is a very different set of figures than when the standards were set beforehand by the ministry. The advantage of the first approach is that the judges have set a standard which they believe should be achieved. The performance results are then matched against this standard. Even if the standard is not reached the ministry knows how far off the standard its pupils are. The second approach has the disadvantage that it creates a standard, or a number of standards, according to the existing distribution of scores, but none of which may reach the standard set by the first approach. In this sense, the ministry receives information that will not help it to judge the gap between what it wants and the existing scores.

Which items constitute a score? Standards of achievement are, of course, a tricky matter and there are often disagreements about what such standards should be. It is not an easy matter to reach agreement conceptually on what should be measured and which items can legitimately be put together to form a score. It is true that statistical checks can be carried out to see if the items selected can be compiled into a score, but it is the step before the statistical checks that is most important. A few comments will make this clear.

It recently came to the author's notice that in England one of the examining boards regarded the following as meeting the required standard for communication in spoken German:

"Wenn geht die Zug von Hamburg zu Lüneburg?"

The correct German is:

"Wann fährt der Zug von Hamburg nach Lüneburg?"

Thus, in the sentence regarded as meeting the standard in England there were four errors. However, the English maintained that a young

German being asked this question in Hamburg would understand it, and that is the purpose of communication.

This is a rather far-fetched example but it exemplifies the types of difficulties when arriving at the measures to be used for test construction in some subject areas.

The Assessment Performance Unit in England conducted this kind of work until its demise at the end of the 1980s. A new body, the Schools Examinations and Assessment Unit (SEAC), became responsible for measuring the standards of achievement in core subjects in all schools at certain points in the school system. Mrs. Thatcher (1993) has described the difficulties encountered in the development of a national curriculum and the accompanying testing. All she had wanted was a "basic syllabus for English, Mathematics and Science with simple tests to show what pupils know" (p. 593). Indeed, one apocryphal story has it that she said that all she wanted was a basic syllabus in the three Rs (reading, writing and 'rithmetic) and then said "Perhaps I mean the five R's". "What are the other two", asked a member of her cabinet. "Right and wrong, of course", she replied!! The purpose of the testing, in Mrs. Thatcher's eyes, "was not to measure merit but knowledge and the capacity to apply it" (p. 593). As it turned out, the teachers rebelled against the amount of testing and the amount of assessment to be carried out by the teachers themselves. The amount of testing was then decreased and the emphasis was on external assessment. It is clear that the officers of the Department of Education and Science (DES), as it was then known, turned what should have been a fairly simple exercise into an unwieldy juggernaut. The experts employed for the establishment of a core curriculum had difficulty in agreeing on content and were "incapable of distinguishing the wood from the trees".

The Ministry of Education and Culture in Indonesia, on the other hand, took the view that it should have periodic surveys of achievement in key subject areas at grades 6, 9, and 12 levels in the school system. The ministry's administrators were responsible for 27 provinces and several millions of pupils in school. Data collection was a massive undertaking given the very large distances to be traveled by the data collectors. Since 1976 (Moegiadi & Elley, 1979) they have run these surveys. They were very much interested in the differences in achievement between provinces and between schools, between urban and rural areas, between different school types, and between children experiencing different conditions of schooling. They measured many of the major objectives of learning set down in the national curriculum and used the results of the survey to identify those objectives that were being poorly achieved. They subsequently requested the National Curriculum Centre to revise the learning materials dealing with those objectives. In Indonesia, as in many developing countries, the rho (measuring the variance between and within schools) is 0.3 at grade 6, 0.5 at grade 9, and 0.7 at grade 12. The larger the rho the greater is the number of schools needed in the sample drawn because schools are very different. This makes surveys very expensive. This can be compared with the rho in the Nordic countries which is usually slightly under 0.1, thus requiring a much smaller number of schools in the sample for the same precision of estimates of mean values, percentages, or proportions. It should also be pointed out that where there is a highly differentiated secondary school system, as in Germany and the Netherlands, the rho tends to be in the region of 0.5 (see OECD, 1993, p. 158).

Many countries have the same policy as Indonesia. That is, they undertake surveys periodically but not at regular intervals. Some participate in international surveys and add to the test items to make the tests fully valid for the national curriculum. In this way, they gain the advantages of an international survey and a national survey. More will be said about international surveys later in this chapter.

To mention the Zimbabwe case yet again, the reason for measuring the inputs and outcomes was that, on the one hand, there was a desire to learn about the level and equity of inputs but, on the other hand, there was a desire to relate the variation of inputs to the variation in achievement in order to determine which of the many inputs on which data were collected were most related to achievement or, otherwise said, most "influenced" achievement. Resources in the ministry were very tight and the ministry officers knew that it could not improve the levels of all inputs. But, if the more important inputs could be discovered then this is where one could begin. The details of this exercise are reported in Ross and Postlethwaite (1992) and Moyo, Machingaidzi, Murimba, Ntembi, and Pfukani (1994). The latter book is an example of how studies such as the one reported can lead to suggestions to the ministry for direct action.

How Ministries Organize their Data Collections

As already mentioned, there are two major ways in which ministries organize their data collections. They are through the ministry's school census unit (or through a national bureau of statistics) and through national research institutes or university units specializing in survey research. In many countries the ministry's own units conduct both the school census work and the sample survey work.

Censuses and Surveys

At several junctures in this chapter, reference has been made to census and sample survey data collections. A census is where information is collected from all schools in the country. The census unit constructs a questionnaire requesting different sorts of information from each school. Either the school principal or a senior member of staff completes the questionnaire, and it is returned to a central point where the data are entered onto a file and then analyzed and a report prepared for the Minister. If information is required from every school, then it is wise to conduct a census. Even in cases where another agency conducts the census, it is usually a unit within the ministry that develops the questionnaire and gives it to the other agency to collect the data. A sample survey is a form of data collection carried out on a probability sample of schools and pupils within schools. The sample is drawn in such a way that it is possible to make generalizations from the sample to the full population under consideration with a great degree of accuracy. Sample surveys are cheaper to carry out and where a general picture for the whole country is required of the indicators being measured it is usual that the ministry opts for a sample survey. Often it is sample surveys that are used to measure achievement and indicators believed to be related to achievement. In all data collections the measures to be used must be carefully developed and tried out and revised before being used.

The Ministry's Census Unit

Within each ministry there is usually a small census unit responsible for collecting data from all schools in the country. This is typically done once a year but in some countries it is done two or even three times a year. In a few countries the ministry hands over this responsibility to the National Statistics Bureau which is often responsible for the collection of all sorts of statistics for several ministries. Many ministries collect data on school enrollments by grade 3 level and by sex of pupil. Some, but not all, include the age of the pupils. The types of variables on which data are collected vary considerably from country to country. These data are usually aggregated to the national level and then broken down by several variables such as urban/rural, school type, and region. The ministry then publishes an annual 'Education Statistics' book which contains these results. The most recent results are often presented as the last row in tables containing the same statistics for previous years or for all years since this type of data collection first began. However, there are a few ministries that go further and publish information about school buildings, teachers, school principals, and pupils.

Major changes are beginning to occur in the types of data collected as census data and the way in which such data are stored, analyzed and used. As the price of computer hardware is rapidly decreasing, more and more schools are acquiring computers. Several countries are networking the schools' computers so that at least there is a direct connection between the ministry and each school. This is already occurring in several states in the United States of America. This "connectivity" is revolutionizing both the collection of data about schools and the creation of data bases at the central level, so that any school can send requests to the ministry about curricula, textbooks and similar issues and gain immediate access to information. In Hungary, for example, half of the schools in the country had a computer in 1993 and it was considered not to be very long before all schools would have a computer. The Hungarian Ministry's school census unit is preparing for regular data collections consisting of a series of short questionnaires and checklists to be completed by schools and data fed back at various points in the school year. A special Data Entry Manager program is being prepared for the entry of these data so that it is easy for the person at the school level to enter the data accurately. The data will then be aggregated to the municipality level, the county level, the regional level, and the national level. The data being collected will be much more comprehensive than hithertofore, but not all of it will be published. Discussions are being held with the likely users of the data to distinguish between what will be published that is of interest to the public, to the regional officers and to the national-level civil servants, and the way in which this is best presented. Only a fraction of the data entered into the relational data bank will be published. The rest will be stored in such a way that it should be possible for analysis to be carried out in minutes (rather than days) for specific purposes, including the data required for answers in Parliament.

Given this "connectivity" many ministries are rethinking which data should be collected, how the data should be managed, which kinds of analyses should be conducted, and which results at which level of aggregation should be reported to whom, and for which reasons. Or, to put it another way, the whole notion of Educational Management Information Systems is being rethought, renewed, and improved.

National Research Institutes and University Survey Units

Where census data are not collected and where it is not a research unit or planning unit in a ministry that is responsible for the sample survey work, the ministry contracts the sample survey work out to other units. In some countries there is an autonomous or semi-autonomous institute for educational research. Australia, for example, has the Australian Council for Educational Research (ACER), New Zealand, the New Zealand Council for Educational Research (NZCER), although in New Zealand the Ministry of Education prefers to have the sample survey work conducted by its own Research and Statistics Division, Scotland has the Scottish Council for Research in Education (SCRE), and England and Wales have the National Foundation for Educational Research (NFER). It is to such institutions that a ministry often contracts out such work.

In other countries, there are special nongovernmental institutions specializing in sampling surveys. For example, in the United States of America, the governmental body responsible for such research work, the National Center for Education Statistics (NCES) subcontracts the work out to bodies such as Educational Testing Service (ETS), Weststat in the Washington beltway, or the Research Triangle Institute in North Carolina.

In yet other countries there are institutions connected with a university that are asked to undertake the sample surveys of schools and students. For example, the Institute of Educational Research at the University of Jyväskylä in Finland is often asked by the Finnish Ministry of Education to undertake educational surveys, or the Universities of Umeå and Stockholm in Sweden, or the Reading Research Centre in Stavanger in Norway (for a sample survey study of reading achievement). In Israel, the Department of Education at the University of Tel Aviv or the Henrietta Szold Institute are asked to conduct such work. The Dutch Ministry of Education subcontracts school surveys to one of the four regional centers for educational research in Amsterdam, Enschede, Groningen, and Nijmegen. Although the Max-Planck Institute for Human Development and Education in Berlin can be considered to be a national institute for educational research in Germany, the Federal Ministry for Education and Science subcontracts sample surveys of student achievement as part of international research studies to university units.

In summary, there are different types of institutions that are used by ministries of education to undertake monitoring and evaluation exercises. In most countries—especially in so-called developing countries—the work is conducted by ministry units. In other countries there is a mixed picture with the work sometimes being conducted by the ministry units, sometimes by autonomous or semi-autonomous research institutes, and sometimes by units within university departments of education.

International Studies

Chapter 1 has described the evolution of the perceived need for international studies. In the final section of this chapter a few comments will be made about selected aspects of the conduct of such studies. These will include the comparison of like with like, 'fair' international tests, and the way in which analyses are undertaken.

Although there are many countries that do not undertake national monitoring and evaluation studies on a regular basis, there are those countries that 'piggy-back' onto international studies that are run on an increasingly regular basis, especially in the areas of mother tongue, mathematics and science achievement. Three major questions are frequently asked about such studies: "Are like children being compared? Are the tests equally valid for all countries? Do the analyses produce results that are useful to ministries of education for their own use?"

The Comparison of Like With Like

The IEA studies, at their inception, took the view that it was wise to compare age groups. Age was comparable. In its first major study of mathematics achievement (Husén, 1967), two major target populations were selected for study. The first was '13-year-olds' and the second was all pupils in the final grade of the 'academic' secondary school. The first population was age-defined because this group was still in full-time schooling in all of the countries at that time. The second was a grade-defined population. For the first population, pupils were spread over several grades and it meant that pupils from all of these grades had to have a specified non-zero chance of entering the sample. This was no easy task in those countries where there was a good deal of grade repeating. A compromise had to be struck and the target population became "all 13 year-old pupils in the grade where most 13-year-olds were to be found as well as the 13-year-olds in the grade on either side of that grade". In effect, this captured nearly all 13-year-olds. However, it was a lot of work for those conducting the study in each country and for the schools in the sample. It was said that school principals objected and the researchers in each country pleaded for grade-defined populations. However, this resulted in anomalies. The target populations were usually defined as the "grade where most pupils of age X were to be found on a certain date in the school year". But, those responsible for the data collection in the countries continued to object because this still meant drawing a random sample of all pupils within the grade, which caused trouble in the schools. Slowly, the practice emerged of drawing only one class within the grade. This, it was said, ensured school cooperation because it did not cause very much disruption in the schools.

There were two consequences of this approach. First, the target population resulted in very different ages being compared. There could be as much as nearly one year difference in age between the target populations being compared; and, it was impossible to make a correction for age because there was no knowledge of how the same age pupils in other grades were achieving. Second, if only one class per school was selected, then there were no estimates of the grade as a whole, especially where there were substantial differences between the classes within a grade, so that not even a whole grade was being compared. Furthermore, where estimates were made of between-school and within-school variances, the between-school variance would be overestimated where the within-school variance had been constrained because of only one class being selected.

The ministries of education wanted a good estimate of how their systems were 'doing' compared with other systems and one wonders if the comparisons of one grade level with different ages is the best way of achieving this. Perhaps it is time to return to age-defined populations with more effort being made to collect such data. Given that more and more systems have compulsory education until 16 years of age, consideration might also be given to testing that population rather than the 13 or 14 year-old population. The argument advanced for not doing this is that since earlier studies tested 13 and 14 year-olds then it would not be possible to measure change in achievement over time of those populations if one changed the age group. It might be possible to do both. If this is not the case then ministries would have to decide for the old target populations or a new one at the end of compulsory schooling.

Once the target populations have been defined in terms of "desired" target populations, there follows the operationalization of the desired population in terms of the "defined" population. Some countries wish to omit all pupils in special education schools and isolated schools and those who have a different mother tongue than the language of instruction and often other identifiable minority groups. Some standard needs to be set about the amount of the desired population that can be omitted for the resultant defined populations to be comparable. Is 10 percent reasonable?

There are then the technical problems of drawing a sample and the response rates. Again, there need to be some technical standards. It would seem that the organizations undertaking international research should agree a set of standards with the ministries. Or, an international agency such as OECD or UNESCO should undertake this task. And once the standards have been set, they should be adhered to. That is, any country not meeting the stated standards would not have their data reported in international reports or would have their data flagged to indicate that there were some technical imperfections debarring comparison or certain comparisons.

'Fair' Tests

Do children have to learn more or less the same content in mathematics and science by a certain age? If this were so then it should be possible to produce a test that measures the bulk of the curriculum of all nations. The first task is to discover to what extent the curriculum to be learned by a certain point in the school system is or is not the same. This is typically done by means of content analysis of the syllabi and textbooks used in the country. On the basis of the national blueprints of what has to be learned, an international test blueprint is produced. At that point, it is a relatively easy matter to compare the international blueprint with the national content analyses. It should be possible to state the extent (in percentage terms) that the international blueprint agrees with each and every national blueprint. There will never be 100 percent agreement, and the question is what percentage may be deemede reasonable. There seems to be agreement that 80 percent would be reasonable. But, what if one country has 78 percent or 60 percent agreement with the international blueprint? That is, is the ministry willing to participate in a study where the other systems being compared all have at least 80 percent of their curriculum included in the international test and its country only has 60 percent of its curriculum covered by the international test? It is for the ministry to decide this question but this author would be dubious. At the same time, it is for the international researchers to describe fully the curriculum coverage of the international test for each system. This implies a two stage participation in such studies. Part One would be for the curriculum content analysis, and Part Two would only be entered into if there was evidence that 80 percent of the curriculum was covered. Any international publication should also describe this coverage in detail.

If a ministry is prepared to accept 80 percent coverage of their curriculum in a specific subject matter as reasonable for the comparison of the achievement of their pupils with that of other countries, then that is its decision. It was always taken for granted in the IEA studies that systems would add items to the tests to cover the other 20 percent of the curriculum. A few systems did this but unfortunately most did not and were content to "manage" with the 80 percent. The reason for this is not clear but it is likely that the sheer amount of work required to participate in only the international part of the study was too exhausting.

The format of the questions in the test is not an easy matter. It must be recalled that data collection activities of this kind are very expensive and there is a desire to control costs. It must also be remembered that the instrument should use items that test the particular educational objectives as a kind of "summative" evaluation. The test items are not meant to be teaching devices or diagnostic measures. There has been a movement in recent years to demand "authentic" evaluation. This has involved items that have required individual scoring by teams of scorers. The cost of this type of scoring proved to be about five times more expensive than machine scoring. There are studies that show that there is a very high correlation between the two types of items and there are other studies showing a low correlation. It is difficult at this point to see what the upshot will be. In the meantime, it is probably a matter of having agreement among all of the researchers in a particular international study that one or two item types are acceptable. However, this must be a consensus and not one country alone insisting on the inclusion of very expensive items.

Types of Data Analyses

In all of the IEA studies conducted thus far, a large number of independent variables (or indicators) about the pupils, their teachers and their schools were collected at the same time as the pupil achievement was measured. At the beginning of the IEA studies nearly all data analyses were within-system, and they were repeated for all countries, a sort of replication study (cf. Husén, 1967). Where the relationship of an independent variable with achievement was significant (after particular sets of variables had been controlled for) in all or almost all countries then the researchers had confidence in generalizing and stating that "variable X is always or nearly always related to pupil achievement". Where there was no significant relationship in all countries, then this was stated and was regarded as a sort of demythologization of educational speculation (theory would be too strong a word to use). Where there were uneven relationships then the challenge was to try and interpret why the different patterns occurred. In later studies there was a tendency to highlight the inter-system analyses and not have thorough, replicated within-system analyses. Thus, for example, the IEA Reading Literacy study analyzed the differences between all 7,800 classrooms in the study in a massive between-class, pooled analysis (Elley, 1994). But there were no replicated analyses. This was because the time schedule and constrained budget did not allow the replicated analyses to be undertaken even though the researchers involved in the study wished to undertake both the between-system analyses and the replicated analyses. These analyses were left to secondary data analysts who could use the data archive.

The question arises as to which types of analyses the ministries in the different countries wish to have or which priority of analyses they wish to have. Or, if they are unable to make statements in terms of analyses then what are the major questions that they wish to have answered. It is the ministries in the countries that have to fund such studies and one wonders if there should not be a more systematic effort made with ministries to have them state their questions for a study before the study begins.

Conclusion

Chapter 1 described the growth of the monitoring movement. This chapter has reviewed existing practices on which aspects of education (inputs, processes, and outcomes) many ministries monitor. Particular emphasis was placed on the monitoring of inputs because this is a problem for many countries.

The way in which ministries organize their data collections are many whether it be through the use of the census approach or the sample survey approach. Some of the major forms of organization have been reviewed. Finally, a few comments were made on how international comparisons might be improved.

The most challenging "innovation" is the introduction of computers into schools, regional educational offices, and ministries of education, and the connections among these three levels of the education enterprise. This permanent link has many implications for the way in which data can be collected from the schools, the frequency with which data can be collected, and the many types of variables that can be collected. No longer will it be necessary to have a once-a-year data collection, since the information can be collected at various points in the school year. The types of data collected can be expanded and the problem of missing data minimized. The greatest challenge, however, is for the policymakers and educational planners to reconceptualize which data they require for which purposes and at which levels of aggregation—to be used by whom, and how.

References

Biggs, J.B., & Collis, K. (1982). *Evaluating the quality of learning: The SOLO taxonomy.* New York: Academic Press.

Bloom, B.S. (Ed.) (1956). *Taxonomy of educational objectives: The classification of educational goals,* Handbook 1: *Cognitive domain.* New York: McKay.

Bosker, R., Creemers, B.P.M., & Scheerens, J. (in press). Conceptual and methodological advances in empirical educational effectiveness research. *International Journal of Educational Research* (special issue). Oxford: Pergamon Press.

Elley, W.B. (1992). *How in the world do students read?* The Hague: International Association for the Evaluation of Educational Achievement.

Elley, W.B. (Ed.) (1994). *The IEA study of reading literacy: Achievement and instruction in 32 school systems.* Oxford: Pergamon Press.

Husén, T. (Ed.) (1967). *International study of achievement in mathematics: A comparison of twelve countries. Vols. 1 and 2.* Stockholm: Almqvist & Wiksell.

Johnstone, J.N. (1981). *Indicators of educational systems.* London: Kogan Page.

Moegiadi, M., & Elley, W.B. (1978). Evaluation of achievement in the Indonesian education system. *Evaluation in Education 2,* 281-351.

Moyo, G., Machingaidzi, T., Murimba, S., Ntembi, R., & Pfukani, P. (1994). *The analysis of educational research data for policy development: An example from Zimbabwe.* Paris: International Institute for Educational Planning, UNESCO.

National Academy of Education (1993). *Setting performance standards for student achievement.* Stanford: US National Academy of Education, Stanford University.

OECD (1993). *Education at a glance: OECD indicators.* Paris: Center for Educational Research and Innovation, OECD.

Postlethwaite, T.N. (1967). *School organisation and student achievement: A study based on achievement in mathematics in twelve countries.* Stockholm: Almqvist & Wiksell.

Postlethwaite, T.N. (1994). Calculation and interpretation of between and within school variation in student achievement (rho). In G. Phillips (Ed.), *Indicators of Education Outcomes*. Paris: OECD.

Postlethwaite, T.N., & Ross, K.N. (1992). *Effective schools in reading: Implications for educational planners*. The Hague: International Association for the Evaluation of Educational Achievement.

Ross, K.N., & Postlethwaite, T.N. (1992). *Indicators of the quality of education: A summary of a national study of primary schools in Zimbabwe*. Paris: International Institute for Educational Planning, UNESCO.

Schmidt, W.H. (1993a). Measuring learning opportunities and instructional practices in mathematics and science in large-scale surveys: Examining the state of the art. Discussion paper read at the annual meeting of the American Educational Research Association (April, 1993). Atlanta, Georgia.

Schmidt, W.H. (1993b). The Development of a model for the international study of learning ppportunities. Discussion paper read at the Midwest Regional Meetings of the Comparative and International Education Society (October, 1993). East Lansing, Michigan.

Silanda, E., & Tuijnman, A.C. (1989). Regional variations in the financing of primary education in Zambia. *International Journal of Educational Development 9(1)*, 5-18.

Thatcher, M. (1993). *The Downing Street Years*. London: Harper Collins Publishers.

Wolf, R.M. (1993). The national assessment of educational progress: The nation's report card. In *Assessment Literacy: Part 2: Evaluating school effectiveness and setting standards for students*. NASSP Bulletin 77(566), 36-45.

Chapter 3

The Design of Indicator Systems*

NORBERTO BOTTANI and ALBERT TUIJNMAN

Center for Educational Research and Innovation, Organization for Economic Cooperation and Development, Paris, France

Characteristics of education indicator systems, and principles and frameworks for their design, are discussed in this chapter. It also considers the question why countries appear to be interested in comparing their education systems using indicator information. Several examples of indicators published by the OECD and UNESCO are presented to illustrate the importance of consistent definitions and procedures for the calculation of indicators. Examples are also given of several country-level indicator sets. These show that the framework proposed by the OECD can be useful to countries seeking to develop their own sets of indicators for the monitoring of educational progress. To be useful as a management tool in policy analysis, knowledge is needed about the inter-relationships among the indicators in the system. The chapter offers some suggestions for research in this area.

The purpose of this chapter is to present principles for the design of indicator systems and to illustrate their variations in several industrialized countries. The first section introduces the concepts of indicators and indicator systems, and presents a framework for their organization. The second section seeks tentative answers to the crucial question why countries are interested in developing indicator systems and comparing their educational performance. It also addresses the reasons why the Organization for Economic Cooperation and Development (OECD), which is essentially an economic organization, has taken the lead in developing education indicators for international comparisons. Despite the gaps in the information system, and although

---

*The views expressed in this chapter are those of the authors. They do not commit the Organization for Economic Cooperation and Development or its Member countries.

many indicators are beset by both conceptual and methodological shortcomings, the OECD framework of education indicators is considered the most comprehensive one of those that have been developed to date. By way of illustration, some of the indicators published in OECD's indicator report, *Education at a Glance II* (OECD, 1993a) are compared with those offered in UNESCO's *World Report on Education* (UNESCO, 1993). The examples given in this section show that there are no hard-and-fast rules about the kinds of statistics that may qualify as indicators. A further section gives examples of how several countries have added to OECD's indicator framework (see Fig. 3.1), and section five summarizes the challenges ahead. It is considered that insufficient attention has so far been paid to the analytical perspective in the design of indicator systems. Questions such as how the indicators are interrelated and what predictive power they have are rarely addressed systematically. As an illustration of what might be done, the last section presents estimates of the correlations among a small sub-set of education indicators. The results are not definitive; they are presented only in the form of hypotheses for further investigation.

Conceptual Framework

A few salient aspects of indicators in general, and education indicators more particularly, are described below. Attention is also given to how education indicators can be organized into a system, and how they can be used.

The Indicator Concept

The definition of an education indicator must be considered first. Such an indicator is generally defined as a policy-relevant statistic designed to provide information about the condition, the stability or change, the functioning, or the performance of an education system or a part thereof. Education indicators may point to—or index—a certain aspect of the condition, functioning, or performance of an education system, but they are not the phenomenon itself.

A litmus test can show a change in color in the presence of a chemical substance. An indicator may similarly show a state or change in education, as in the case of meeting a desired standard of performance or a danger signal of relative inefficiency. Quite simply, an education indicator can act as a warning device that something might be going wrong, much as the instruments on the dashboard of a car warn drivers of malfunctioning equipment and operating systems, or to reassure them that all is well. If a red light comes on, say the oil indicator, then drivers know that they had better stop and check the engine. However, such an indicator does not tell precisely what went wrong or what must be done to fix it.

An economic, social or education indicator is a single or complex statistic that conveys essential information about the "state" of an important aspect of a social or economic system, whether this be the economy or the labor market, health, or education. Such indicators function in much the same way as the warning lights on a dashboard. However, they rarely convey information of the type "beware: the light is on" or "rest assured: the light is off". An education indicator presents information on a scale. It is not normally possible to find an unambiguous cut-off point on such a scale. Hence it is not possible to state precisely what point—what indicator value—signifies "acceptable" as opposed to "insufficient" performance. An indicator is therefore not a precise instrument for policy evaluation. Interpretations of the indicator values that signify danger and that suggest a need for policy action, will differ depending on the perspective of the examiner. An indicator, therefore, cannot in itself offer a diagnosis of a problem, nor suggest what can be done to remedy it. Indicators, then, have their limitations, and their utility depends in part on their reliability and validity. These technical qualities of indicators are discussed in subsequent chapters.

Characteristics of Indicators

Indicators are among the tools available to policymakers. They have the following characteristics:

1. They are *quantitative,* but they are more than simply a numerical expression or a composite statistic;
2. They are intended to *convey summary information* about an important aspect of the functioning or performance of the economy or an education system;
3. They are intended to *enlighten and inform* the stakeholders and other interested parties. In the case of education, the stakeholders range from the students and their parents, teachers and school principals, school inspectors, local administrators, employers, and of course politicians and decision-makers in government agencies;
4. They are intended as *diagnostic tools,* as a basis for evaluation, and for creating new visions and expectations;
5. Ideally, they should be *part of a larger set* that includes pointers suggesting how the indicators might be interrelated. Although an indicator alone can be informative, value added can be achieved if knowledge about the relationships among the various economic and education factors is available;
6. They involve, or call for, *value judgments* and they are therefore intimately related to questions of policy. It is perhaps for this reason that indicators often attract much attention from the mass media the world over, precisely because they derive their meaning in a particular political context.

Thus, in conclusion, education indicators are statistics which tell something about the functioning or performance of education systems. They therefore have an inherently political function (Cohen & Spillane, 1994; Darling-Hammond, 1994). This partly explains why not all education statistics qualify as indicators. Nuttall (1992) put it as follows:

Like economic and health indicators (e.g., GNP, infant mortality rates), education indicators focus on key aspects of their topics. Consequently, indicators are expected to reflect the condition of the system as a whole, or of some significant part or element of the system. To be an indicator, an education statistic must also have a reference point against which it can be judged. Usually the reference point is some socially agreed-upon standard (e.g., a minimum reading age to indicate basic literacy), a past value (e.g., the 1970 level of mathematical attainment), or a comparison across schools, regions or nations. Obviously, indicators do not tell everything about education systems. Instead, like economic or health indicators, they provide an "at a glance" profile of current conditions (Nuttall, 1992, p. 14).

What Should be Measured?

This chapter focuses on *system level* indicators, that is, the indicators in which the unit of analysis is a system of national education. Excluded are those indicators in which the unit of analysis is the state, province, city, school, or student. These different units of analysis not only need different types of indicators but also imply specific statistical and methodological issues.

As noted in Chapter 4, it is difficult to define education systems and therefore the phenomena to be measured. A major problem is to select the information that reveals the most important things about an education system. Differences in points of view on this issue are large and concern the problem of what to measure more than how to measure. Current collections of measures of the education system and its processes are heterogenous and diverse because there is no clear and universally recognized concept of the education system. In some cases, the conceptual approach to education is limited to the formal types of public education at the primary and secondary levels only; in other cases it may include private, tertiary education and even adult education. In some circumstances, it is possible to cover a part of private education—at least the private institutions that receive public subsidies—but it is still very difficult to obtain a comprehensive view. Still more difficult to conceptualize and measure are the value-foregone opportunities of parents and students who contribute their time, attention, and energy to the educational process. Additional measures of importance—many of which are unfortunately lacking from the data collections that are undertaken regularly by education ministries and national statistical offices—are mentioned in Chapter 2. The comparative knowledge base currently available thus provides only a partial view of the education effort of individuals and nations.

Systems Indicators

The definition given previously suggests that, other things being equal, a systematic collection of indicators is more valuable than a single indicator. Such a collection may require breadth of scope, balance, and concision (or the omission of unnecessary redundancy). Much judgment is called for meeting these criteria by international and national agencies seeking to satisfy their distinctive purposes.

A wide range of conceptual models for education indicators is reviewed by van Herpen (1992). These models testify to the differing perceptions and purposes of both system models and indicators. What the different approaches have in common, however, is a concern with the relationships between sets of related concepts. Many authors, including Johnstone (1981), Burstein, Oakes and Guiton (1992) and van Herpen (1992), have stressed that no single conceptual scheme can usefully include all important aspects of an education system, because education is multifaceted and multidimensional. Consequently, a unified perspective is untenable, and an accepted framework for organizing the education indicators and depicting their complex interrelationships does not exist.

An Organizing Framework

Many epistemological and practical issues arise in the construction of an indicator framework. Realizing this, the indicators the OECD published in *Education at a Glance I* (OECD, 1992a) were selected for both conceptual and pragmatic reasons. Some of the indicators derived from logical relations among the various parts of the education systems and had an empirical orientation, whereas others were the result of practical concerns and were policy-sensitive in their orientation. The advantage of this combined approach was that "it strikes a balance between stability and flexibility in a set of indicators that will evolve with time" (OECD, 1992a, p. 11).

The OECD indicators were moreover organized into a framework that was suggestive of relationships and the presence of a sequential ordering among the indicators, without, however, calling attention to predictive, cause-and-effect interpretations. The decision to depict the organizing framework in this way was not only influenced by political considerations. It was also a consequence of the disagreement over the notion of causality among academic researchers representing various disciplines and—at times apparently irreconcilable—traditions in the epistomology and methodology of the social sciences (OECD, 1992b).

The framework employed for the set of 38 indicators published in *Education at a Glance II* (OECD, 1993a) is shown in Fig. 3.1. It has three principal divisions: context information, input and process indicators, and outcome indicators.

Education systems and the results they produce do not exist in a vacuum; they are the product of a complex historical process, and are influenced by many factors in the surrounding environment. While some of these factors are malleable and can be optimized, many others are given and cannot be changed through education policy. An analysis of education must therefore be informed by an appreciation of the educational processes employed, and the financial and other resources expended, against a background of contextual factors in the environment of education systems. There are many factors that condition the flow of financial and human resources available for education, and that influence the learning culture of a country. Contextual indicators might be of a demographic nature covering aspects such as the proportion of school-aged children in a population, the percentage of students who are members of disadvantaged minorities or the distribution of students by predominantly rural or urban areas; they might also be primarily economic, for example, the wealth of a country as measured by its gross domestic product or the proportion of school-aged children living in poverty; or they could have a social and cultural orientation, for example measures of community resources that can be relevant to education, or indicators of the social and economic conditions of the students' homes, the educational careers of their parents, or whether the students are proficient in the language used in school.

Examples of the contextual indicators included in Fig. 3.1 are the percentage of the population that has received a certain highest level of formal educational attainment, employment and unemployment ratios, and an index of gender inequality in educational attainment. The decision about which indicators to include in this first cluster of the framework is not an arbitrary one. Restrictions are imposed because an indicator set should not be overloaded with information. Other limitations arise because the range of statistical data required for measuring a wide spectrum of context variables is not currently available at the system level. Future editions of OECD's indicators report can hopefully draw from an increasing variety of available context information.

The indicators mentioned in the second cluster of the framework are grouped into four sections: finance, human resources, participation, and decision-making in education. The eight finance indicators compare the educational expenditures of the countries from a variety of perspectives. The two indicators on human resources offer information about the size of the education sector relative to the entire labor market and about staffing patterns within the education sector. The seven participation indicators offer information on enrollments in the education systems of the countries. They focus on early childhood education, upper secondary education, and tertiary education—all levels in which enrollment has increased but is not universal. The four indicators on the locus and modes of decision-making in education present the results from a special survey conducted under the auspices of the OECD in 1992. The purpose of the survey was to discover how and by whom the various decisions affecting education systems were made.

The outcomes of education can be examined and interpreted in different ways. Regardless of which results and related characteristics are considered, they always reflect both the historical conditions and the combined effects of policies, programs, practices, and decisions that contribute to the education system in each country. The outcomes of education, however defined and measured, are thus a function of a complex set of social, cultural and economic factors, of which education inputs such as money spent, teachers employed and the number of students admitted are but one, albeit crucial, part.

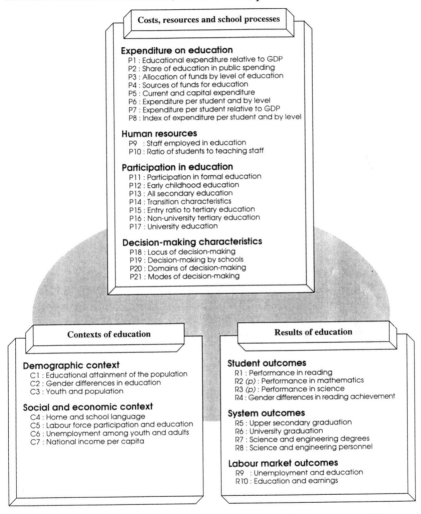

Figure 3.1. Organizing framework for 38 education indicators proposed by the OECD.

The last cluster in the framework includes 10 indicators that show different aspects of the performance of education systems. They are divided into three sections, concerned respectively with student outcomes, system outcomes, and labor market outcomes. The orientation and substantive meaning of the indicators in each section are obviously different. In *Education at a Glance II* (OECD, 1993a), student-level outcomes are assessed by performance in reading literacy, mathematics, and science; system outcomes by the graduation rates in upper secondary and university education; and labor market outcomes by the relationship between education, on the one hand, and employment, unemployment, and earnings from work on the other.

Why Use Indicators to Compare Education Systems?

Why should one want to compare education systems? Embedded as they are in history and culture, they are inevitably so different. But these differences can also be revealing, since they show that there is no single solution, no standard model of education. The insights gleaned from international comparisons are a necessary part of the knowledge base needed for the monitoring and evaluation of the education system, for guiding and developing policy, and for informing decision-making.

The interest of many countries in what occurs in education outside their borders reflects the belief that comparisons can offer information useful for deciding if, and where, educational reform may be needed. In turn, the desire to reform and improve education hinges on perceptions about the economic importance of education and the belief that education is a productive and long-term investment.

Importance of Education in the Economy

To be useful at the system level, a set of education indicators should offer a valid view of the full scope of education in the different national systems. The part of public educational expenditure relative to the gross domestic product (GDP), for example, can be considered an essential indicator that gives an immediate idea about the magnitude of the relative "effort" to support education. An indicator measuring the level of educational expenditure relative to GDP is useful for comparing the shares of economic resources that countries devote to education. The economic scope of education in the industrialized countries can be inferred from the size of the enterprise: In 1991, about 6.4 percent of the collective GDP of the OECD countries was channeled into education, a proportion that was more important than that of the agriculture sector. In several respects this figure is an underestimate; for many countries it includes no data on private expenditures; moreover, it includes no estimates of the cost of parental time spent in the first six years of their childrens' life; nor does it include the continuing value of parental time during the years the children attend primary, secondary, and tertiary education institutions. The estimates furthermore lack additions for the value of students' time, particularly the foregone opportunities to work beginning in the adolescent years.

The importance of education relative to GDP is confirmed by another simple indicator. In the "typical" OECD country, in 1991, 3.1 percent of the labor force was employed in teaching. Among the countries that reported non-teaching as well as teaching staff, total employment in education (all types of staff combined) typically amounted to 5.1 percent of the labor force. This shows that, on average for the OECD countries, the education sector was a more significant employer than agriculture.

Education at a Glance I-II showed the extent of youth unemployment in many countries. However, the indicators in these two publications also demonstrated the strength of the link between education and work: the higher the level of education reached by men and women in the labor force, the higher their labor force participation rates, the lower their chances of becoming unemployed, and the higher their earnings from work. This is not to say that education is a panacea for solving the unemployment problem. It is unlikely that unemployment would disappear simply because all members of a society attained "highly educated" status. But in reducing the risks of unemployment and possible social or economic exclusion, education certainly does count.

The above mentioned data illustrate the relevance of educational services in economic activity. Yet the education sector has apparently not received the attention it deserves from an economic point of view. Indicators such as the measures previously mentioned could therefore help to modify the perception of education at the political level and its application in macro-economic analysis. The ambition of many countries to increase the quality, equity and efficiency of their education systems can be explained by these figures showing the importance of education in the economy.

International comparisons of education systems are of course not undertaken solely to portray the magnitude of the enterprise. There are other motives, and these concern the general desire to improve the competitiveness of the labor force. This is a key issue in all countries, not least because the skills of the population are believed to determine people's life chances, the productivity of firms, and ultimately the wealth of the nations. In the final analysis, the economic conditions on which jobs and quality of life depend is determined by labor quality and adaptability. In turn, high-quality education and training systems are major determinants in ensuring a competent, adaptable, and productive work force.

Concern with System Performance

All countries are concerned with educational performance. Performance can be measured in various ways, for example, in terms of student achievement in key subjects, graduation rates, or the transition from school to further education and the labor market. Other aspects of performance involve knowledge, judgment, insight, and values, and these are not usually measurable in any precise sense of the term. This is to say, performance has both qualitative and quantitative aspects, and these must be related to the different contexts and the inputs and processes by means of which the education systems operate.

It was noted in Chapter 1 that when the Ministers of Education met at the OECD in Paris in November 1990, the political commitment to raising education standards was as strong as ever. But in addition to this, the

Ministers also underscored the importance of education indicators. They considered that comprehensive and comparable education indicators play a major role in accountability and informed decision-making. In their *press communiqué* (reproduced in OECD, 1992c), the Ministers called for a large effort to strengthen the comparative knowledge base in the field of education. This knowledge base would require not only comparable statistical data—especially student performance data—but also knowledge gained through the periodic reviews of national education systems and analyses of country experience in dealing with particular policy issues, for example, how to cope with the pressure of numbers in mass tertiary education or the need for environmental education.

Concern with Education Standards

Countries are anxious to compare their educational performance by means of indicators because they are concerned with quality and standards. What level of performance is deemed acceptable and which not? Clearly, adequate benchmarks are needed for the setting of *opportunity to learn* standards (mainly the inputs into education), *content* standards (mainly the processes of education), and *performance* standards (the student, system, and labor market outcomes of education). The interest of many countries to compare their education systems derives from the realization that knowledge about the performance levels achieved by other countries can offer useful insights and a point of reference for the benchmarking of the performance standards that should be applied. In the United States, France, Sweden, the Netherlands, and increasingly also in other countries, goals for education have been agreed that make reference to external standards. The ambitious growth targets set by some countries have added a political dimension to education, and they no doubt have led to support for international, comparative studies in education. If one wants to have world-class standards, then one should be able to define what is meant by "world class". One way to do this is by means of comparative data collection and analysis. To be useful, such comparisons must take account of the critical parameters of education systems; comparisons should not be limited only to inputs or only to outcomes.

Concern with Educational Reform

All countries are committed to improving quality and raising the level of education in the population, but where are they to begin? Education indicators and international comparisons can offer some guidance. They can help the various stakeholders to begin to ask the right questions and, by signaling strengths and weaknesses, they can point to areas where effort can be usefully focused. Before a problem can be addressed, it must be recognized as such. Indicators will help do that but they cannot tell how the aims can be achieved. But improving the knowledge base in itself is not the only or foremost objective. The goal is to increase the understanding of how education systems function, so that better policy can be devised and implemented.

So why have countries put such a considerable effort into the development of comparative education statistics? The answer may well be: Precisely because data analyses are needed that help to increase understanding, assist in policymaking, and in the long run can contribute to the goal of economic and social development. But can education indicators make such a contribution?

They certainly offer new insights about meaningful similarities and differences among the education systems of the countries. For example, one indicator in *Education at a Glance II* offered detailed information on the transition from compulsory schooling to upper secondary education. The data revealed clear differences among the countries in the numbers of students who make this transition, with high rates in countries such as Germany, the Netherlands and, until quite recently, relatively low rates in Sweden and the United Kingdom. As a result of the attention by the media, there was a policy response in the latter two countries. Policy reactions were also issued by officials in Italy and the Netherlands. These examples testify to the policy relevance of the vast work program that is being carried out to improve the information system in education at the international level.

Examples of Education Indicators

This section presents some examples of international education indicators. The focus is not on the results—although these should be of interest to many readers—but on the procedures and methods used in the calculations, especially in so far these impact on the validity, reliability, and comparability of the indicators.

As will be explained below, private expenditures must be accounted for in the financial statistics if a comparable and full picture of countries' total expenditure on education is to be obtained. The same problem appears in the enrollment data, which can include several distinct aspects such as the demographic conditions of a country, the size of its school-age population, and the flows of students in the system.

There are no hard-and-fast rules about the kinds of statistics that may qualify as indicators. Indicators may be simple numbers, or they may be complex indices. For example, an indicator of participation in formal education that measures the number of students per 100 persons in the population 5 to 29 years of age who are enrolled at all levels of formal education is a simple, albeit synthetic, ratio that allows for comparisons of gross enrollments ratios in any level of the education system. By showing the ratios separately for full-time and part-time students and for public and private enrollment, these ratios can constitute indicators of the overall magnitude of these sectors. This indicator, which is shown in Figure 3.2, is included in the OECD framework. It offers an insight into the schooling "effort" of the countries surveyed.

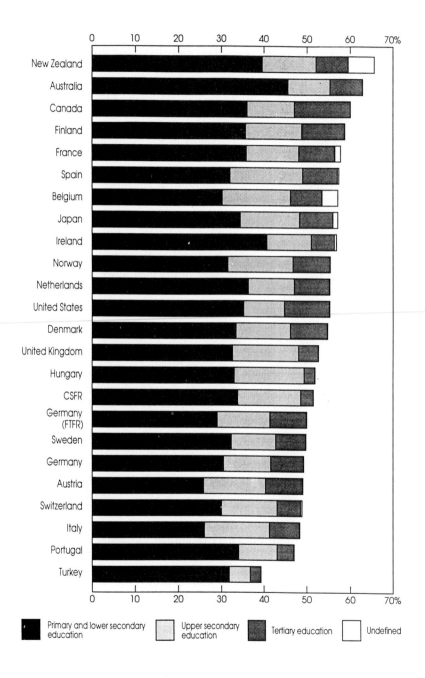

Figure 3.2. Percentage of population 5-29 years who are enrolled in formal education, 1991, Source: OECD (1993a).

Table 3.1. Gross enrollment rates for the population aged 6 to 23 years, 1990

Country	Enrollment rate
Canada	89
United States	86
Finland	84
Spain	83
Germany (FTFR)	81
New Zealand	78
France	78
Denmark	77
Norway	77
Belgium	76
Switzerland	74
Ireland	73
Japan	73
Greece	72
United Kingdom	72
Australia	71
Netherlands	71
Sweden	70
Portugal	66
Italy	64
Austria	64
Turkey	52

Note: Germany (FTFR) refers to the former territory of the Federal Republic of Germany.
Source: UNESCO (1993, pp. 124–127).

Table 3.2. Expected years of schooling of a hypothetical age cohort, 1980
and 1990

	1980	1990
Canada	14.3	16.0
United States	14.6	15.4
Germany (FTFR)	n.a.	15.2
New Zealand	13.4	14.6
Norway	13.1	14.6
France	12.7	14.5
Denmark	13.5	14.5
Belgium	13.5	14.2
Switzerland	13.2	14.1
Netherlands	13.5	13.9
United Kingdom	13.1	13.6
Japan	12.9	13.3
Sweden	12.7	13.3
Greece	12.1	13.2
Australia	12.0	13.1
Ireland	11.3	12.6
Turkey	n.a.	8.9
Austria	11.2	n.a.
Spain	12.5	n.a.

Note: Germany (FTFR) refers to the former territory of the Federal Republic
of Germany.
Source: UNESCO (1993, pp. 124–127).

From Fig. 3.2 it appears that, in 1991, the number of full-time equivalent students enrolled in education (excluding those in early childhood education) per 100 persons in the population 5 to 29 years of age is at least 45 in every country but Turkey. In four countries (Australia, Canada, Finland, and New Zealand), the ratio meets or exceeds 60. The fifth country is France with a ratio of 57.7, but this country as well as New Zealand would—presumably—have been nearer to the top of the list if early childhood education had been included in the calculations.

This hypothesis can be empirically investigated by comparing the results in Fig. 3.2 to those given for the same indicator in *Education at a Glance I* (OECD, 1992a, p. 69), which are based on the 1987-88 school year, and where the calculation took into account the population spanning from 2 to 29 years of age. In the countries with the highest enrollment ratios—Belgium, France, Spain, and the Netherlands—early childhood education was nearly universal for children beginning at the age of three. The mean enrollment ratio for these countries exceeded 60 percent. Hence it can be concluded that the rank order of the countries in this indicator will change dramatically depending on whether early childhood education is included. Indeed, the data reflect two phenomena: the differences in participation across the countries, and the relative size of the specified age groups. For example, participation in primary education will be high if the percentage of a country's population falling within the group designated for primary school attendance is high. Analysts should consider the influence of such definitional and demographic factors on the results when interpreting the education indicators.

In UNESCO's *World Report on Education* (1993), the total enrollment ratio is presented in two ways: The enrollment ratio for the population 6 to 23 years of age, and the estimated number of years of schooling that a student of a certain age can expect. By limiting the age group to the population of 6 to 23 years of age, UNESCO's definition almost entirely excludes the sector of early childhood education and a part of tertiary education. While the chosen reference age group may be more appropriate for comparisons among the Member countries of UNESCO, the consequence is that the classification changes, as one might expect, the ranking of the OECD countries. The results in Table 3.1 are gleaned from UNESCO's *World Report on Education* (1993).

In Table 3.1, almost all the countries have an enrollment ratio higher than 60 percent. The countries ranking at the top of the table all have a high gross enrollment rate at the tertiary level of education. This demonstrates that a change in the numerator can result in a substantially different ranking of the countries.

UNESCO presents a second version of the same participation indicator in the form of expected years of schooling. It is calculated by adding the enrollment rates by age for the first, second and third level. This approach provides useful information and corrects some of the distortion due to the demographic factor. The hypothetical length of schooling of the students who begin their formal education in a given year—calculated on the basis of the assumption that they will meet a stable structure of education and measured using the net enrollment rates age by age—is quite different among the countries. For the OECD countries, UNESCO (1993, pp. 124-127) presents the estimates shown in Table 3.2.

The ranking in Table 3.2 offers an indication of the theoretical duration of schooling in the formal education system. Adult education is excluded from these estimates. The number of years of schooling can be affected by grade repetitions, by the starting age taken in account (e.g., the beginning of compulsory education or the beginning of early childhood education), by the full-time or part-time models of education, and by the age at which young adults complete their formal education. If completion is set at too low an age as is the case for many OECD countries included in UNESCO's indicator, then the values will underestimate the duration of schooling in countries such as Germany, where university studies are of a long duration, or Denmark, Finland, and Sweden, where a large number of young adults consistently begin their university education at a late age.

Table 3.3. Schooling expectancy for the top five and bottom five UNESCO countries, 1990

Country	Schooling expectancy (in years)
Niger	2.1
Burkina Faso	2.4
Guinea	2.7
Djibouti	3.4
Bangladesh	5.0
Canada	16.0
United States	15.4
Germany (FTFR)	15.2
New Zealand	14.6
Norway	14.6

Note: Germany (FTFR) refers to the former territory of the Federal Republic of Germany.
Source: UNESCO (1993, pp. 124–127).

The inclusion or exclusion of early childhood education has major implications for the reliability, validity, and comparability of the participation indicators. This can easily be demonstrated by comparing the results of UNESCO's calculation with those reported by the French Ministry of National Education in its domestic set of education indicators (France, 1993). The French report a value of 18.1 years for the schooling expectancy of a new age cohort in 1990, compared with 14.5 years calculated by UNESCO. This difference is attributable mostly to the fact that France includes early childhood education—which is tuition-free and begins at the age of two—in the calculation.

The indicator on the theoretical duration of schooling allows for relatively straightforward comparisons of the development of schooling across countries. Duration is assumed to be an important factor in literacy and is indicative of the position of schooling in the socialization processes of contemporary societies. The indicator moreover offers an explanation of the apparent discrepancy between the generally observed increase in the number of education staff, on the one hand, and the decrease in student numbers due to the demographic decline, on the other. The increased duration of education compensates for the demographic fall and generates an extended demand for teachers. The indicator can therefore explain why the reduction in the number of enrolled students does not produce an equivalent decrease in the number of education staff. France (1993) reports that despite the demographic fall between 1982 and 1991, the numbers of enrolled students grew by 650,000 as a consequence of an increase in school duration. If the demographic conditions prevailing in 1982 had been stable until 1991, then the number of enrolled students would have increased by 1.3 million. This example shows that synthetic indicators—such as the one on the theoretical duration of schooling—can offer more appropriate analytical opportunities than simple indicators.

The schooling expectancy in France increased by almost 1.5 years in less than a decade. It was 16.7 years in 1982 and 18.3 years in 1991. During this period, the duration of schooling before the age of six did not change. It was 3.2 years in 1982 and 3.3 a decade later. The main change must therefore have occurred in the duration of schooling after 6 years of age. The figures bear this out: It was 13.5 in 1982 and 15.0 in 1991. This means that students stay longer in school than before. In Denmark, during the period from 1982 to 1991, the expected duration of education increased from 13.1 years to 14.1 years—an increase of one year.

UNESCO's figures for schooling expectancy in the top five and the bottom five countries are presented in Table 3.3.

The figures given in Table 3.3. are crude, however, and do not allow for the identification of convergency trends among the countries. In several countries of the South the schooling expectancy at age 5 is more than 10 years: Saudi Arabia (12.2); Bahrain (11.9); Brazil (10.4); Chile (11.7); Cuba (12.7); Unified Arab Emirates (10.6); Jamaica (11.8); Jordan (12.0); Mexico (10.8); Panama (11.1); Peru (11.8); Philippines (10.5); South Korea (13.3); Surinam (11.8); Trinidad and Tobago (10.8); Tunisia (10.3); and Venezuela (10.7). In several OECD countries the schooling expectancy at this age is less than 14 years. This means that the differences in terms of education enrollments and schooling duration between the OECD countries and many developing countries are less than four years and, at least for some countries, less than two years. This shows considerable progress in the expansion of schooling world-wide.

The examples presented above demonstrate the importance of the definition and calculation procedures, especially when indicators are used less for managerial purposes and more for the analysis of general trends and differences among countries. System-level indicators often require a high level of data aggregation. The same data may be used to construct different indicators. Moreover, each indicator is affected by many other factors, some of them directly connected to education. Therefore, much attention should be devoted to the composition of variables and to the organization of the data used for calculating indicators.

National Indicator Systems

Following the publication of *Education at a Glance I* (OECD, 1992a), a number of countries began to create their own, national indicator systems and published documents related to the perspective offered by the international set. The OECD framework was no doubt used as a leverage for producing other comparative reports on education indicators (cf. Nordic Council, 1991). By January 1st, 1994 the following countries were known to have published a national set of education indicators: Belgium, Denmark, France, Sweden, Switzerland, and the United States.

The links between the OECD indicators and the various sets of national indicator systems reviewed below were established by including similar measures in both sets—that is, by employing the same definitions and calculation procedures. The report, *Education in States and Nations: Indicators Comparing US States With the OECD Countries in 1988*, published by the US National Center for Education Statistics (NCES, 1993) replicates 16 key indicators chosen from among those published in *Education at a Glance I*. In the United States, *Education in States and Nations* has been presented as a logical step and companion volume to the OECD reports. As mentioned in Chapter 1, this innovative US report allows not only state-to-state and country-to-country comparisons, but also state-to-country comparisons.

Other countries have adopted a different approach. The French Ministry of National Education, for example, has developed a set of indicators for system monitoring that includes a few OECD indicators (France, 1992a; France, 1992b). The international results are used to anchor the much more detailed French results, and to compare them to those of other countries, most often the ones associated with the so-called "Group of Seven". Another approach was adopted by Switzerland which published a number of education indicators that made it possible to compare the country-level results with those of other—often neighboring—countries, as well as allowing between-canton comparisons. For the Swiss, it was the first time ever that cantons were compared among themselves.

Theoretically, countries such as Australia, Canada and Germany, while gathering data for the international set of indicators, could easily generate a national companion set that employs the same organizational framework and applies the same calculation procedures. A setback may be that the data used at the international level often differ from those used nationally. The various audiences may have difficulty to appreciate the differences that arise from the use of different definitions. Be this as it may, the countries engaged in the development of the international set know the procedures that were applied. They have, moreover, collectively participated in the establishment of the definitions and, in the mid-1970s, in the adoption of the International Standard Classification of Education—(UNESCO, 1992). Furthermore, they receive from the OECD Secretariat the weighted indices emanating from the national accounts framework—GDP, per capita GDP, and the purchasing power parities (PPP)—that are required if duplication of the international results is deemed necessary.

In conclusion, it may be considered that the countries wishing to develop their own national-level sets of indicators can do so by following the model proposed by the OECD. Because such within-country developments are innovative and instructive, they are described in some more detail below.

Belgium (Flemish Community)

The Belgian authorities published a set of indicators adapted to the education system of the Flemish Community (De Groof and Van Haver, 1993). They addressed the needs to examine the overall performance of the education system, to evaluate trends, and to compare the performance of the Flemish system with that of neighboring countries of France, Germany, the Netherlands, and the United Kingdom. In some cases comparisons are also made with other countries, notably Canada, Finland, Spain, and the United States. The report contains 26 indicators divided in three sections: processes, results, and background, and is organized on the basis of the OECD framework shown previously in Fig. 3.1. The international comparisons are established using 23 OECD education indicators. Other data sources assembled by IEA and EUROSTAT, the statistical office of the European Union, are also employed.

The report includes several indicators that were proposed at an early stage by the OECD but that have not yet been calculated at the international level. Examples are an indicator of special education, another on the public expectations and attitudes toward education, and an indicator on teacher remuneration. In the process domain, three indicators offer information about the use of computers in schools based on data collected for the IEA study on computers in education (Pelgrum & Plomp, 1993). The Belgian report also includes data on student achievement employing the multiple comparison charts that were established by the US National Assessment of Educational Progress (NAEP) and used by the OECD in *Education at a Glance I-II.*

Denmark

Produced by the Ministry of Education and Research, the education indicators report of Denmark (1993) includes 37 indicators organized in three sections: resources, activities, and results. The selection of indicators reflects the intention to show the evolution of the Danish education system during the 1980s and to identify the tendencies at the beginning of the 1990s. The indicators generally cover a number of years, most often the period 1981/82-1990/91. The international comparisons between Denmark and the other countries concern the sections on resources and results. Denmark employs seven indicators contributed by the OECD.

A distinctive point of the Danish set is the inclusion of information on enrollment in and expenditure for adult education. Other unique indicators are also included: unit cost per graduate, expenditure for the state education grant and loan scheme, student income, expenditure for research, time spent in education, and waiting time before beginning study. An innovation in the results section is the measure of employment one and five years after graduation.

France

When the French Parliament discussed the education budget in 1993, the Minister of Education presented each delegate with a copy of *L'état de l'école* (The Condition of the School). Two editions of this report have appeared in 1992 and a third in 1993 (France, 1992a; France, 1992b, France, 1993a). Each report contains 30 indicators produced by the statistical branch of the Ministry. The main themes are the analysis of relevant trends and the performance of the education system in an international perspective.

The organizational framework of the 30 indicators is more elaborate than the Danish one: The same three sections are taken into account (e.g., resources, activities, and results) but for each section the indicators are organized by domains: generalities, first level, second level, third level, and adult education and training.

The French authorities have devoted a special section of the introduction to the results emanating from international comparative studies. A systematic link with three countries provides a reference for the assessment of performance in the French education system: Germany, Sweden, and the United States. In order to identify the relative position of the French system, the data analysis considers three indicators: the index of expenditure per student and by level; the enrollment rates at 4 and at 21 years of age; and the relationship between unemployment and education. Moreover, six other OECD indicators are used for comparisons with mostly other European countries.

The most recent report (France, 1993a) includes unique data in each section. Expenditure on adult education and training is considered in the section on resources. In the activities section are expected duration of education, geographical disparities in the rate of graduation from upper secondary education, and the time devoted to different subjects in the intended curriculum. Part of the results section are indicators concerning the educational attainment of conscripts, disparities in graduation rates among the senior high schools, the educational expectations of children, and the relationship between educational qualifications and social conditions. The report also contains some indicators based on an annual assessment of the performance of 5th and 6th graders in two subjects, French and mathematics.

Finally, it is worthwhile mentioning that France (1993b) also publishes a second report on education indicators, entitled *La Géographie de L'Ecole* (The Geography of the School). This set of education indicators compares the performance of 23 geographical areas (called *Académies*) and reports on within-country disparities in education.

Switzerland

Together with the United States, Switzerland is one of the most decentralized of the OECD countries, at least in the education sector. The Swiss federal system involves 26 cantons, each of which administers its own education system. The cultural, linguistic and geographical differences are so great that for years the idea to compare the education systems of the 26 cantons has been inconceivable. The international education indicators proposed by the OECD have helped to modify the resistance to the development of a modern set of national education indicators that allow for both between-canton and between-country comparisons.

Comparability is the key word in the Swiss report, *Les Indicateurs de L'enseignement en Suisse* (Switzerland, 1993). This is understandable, given that this official study marks the first time that the educational performances of the cantons and of Switzerland as a whole are compared to those of other cantons and countries. The latter include the four neighboring countries—Austria, France, Germany, and Italy—plus Japan, the Netherlands, the United Kingdom, and the United States.

The Swiss set of indicators is more encompassing than the one used by the OECD. It includes 45 indicators organized in three sections: contexts, processes, and results. Switzerland made the largest use of the OECD education indicators; no less than 23 international indicators were included in the national set. A number of specific features characterized the Swiss report. In the section on the contexts of education, indicators are included that measure the skills gap in the labor force, the educational background of the students' parents, and the mother tongue of the students. The section on educational processes features an indicator about the typical duration of compulsory schooling (there is a difference of almost two years between the cantons having the longest and shortest duration) and about the cultural composition of school classes. Indicators of the curriculum are unfortunately missing from this section. Finally, a whole cluster of indicators on the participation of adults in further education and in cultural activities is included in the results section.

United States

The United States' effort to collect and disseminate comparative data on education is not new. *The Condition of Education* (NCES, 1992), and the annual publications by the US National Education Goals Panel (NEGP, 1991-1993), are among the visible products of this effort. Both make use of a variety of data sets, among which the NAEP (see Chapters 10 and 11), and provide a rich statistical picture of American education, including trend information, comparisons among various groups (e.g., by sex, ethnicity, and socioeconomic status), and international comparisons. The NEGP collects, organizes and develops education indicators that particularly pertain to the six national goals proclaimed in Charlottesville in 1990 (see Chapter 1).

Both NCES and the NEGP have used the OECD education indicators, but the most systematic and innovative comparison between the US and the other OECD countries is presented in the previously mentioned report, *Education in States and Nations* (NCES, 1993). This report includes 16 indicators grouped in four sections: background, participation, outcomes, and finance. Two background indicators—population and land area—offer useful insights into the differences in size between the US states and the OECD countries. These two indicators have not been considered in the OECD set. All other indicators follow the same calculation procedures as those reported in *Education at a Glance I* (OECD, 1992a).

Some Challenges Ahead

The development of international education indicators may be said to have entered a new phase. The first international publications on education indicators (OECD, 1992a; OECD, 1993a) were hailed as innovative by some countries, but accepted only reluctantly by others. The work continues, however, and as more precedents are set an increasing number of countries may be expected to begin serious work on the building of a coherent information system for the monitoring of educational progress.

The challenges are of an instrumental, a technical, and a logistical nature. In the initial phases of the work, much of the discussion focused on what then was labelled "the uncertain connection" between indicators and educational improvement. It is true that the OECD indicators carry a certain utilitarian, if not instrumental perspective—the value of the indicators will only be proved by their usage. The real test of the indicators is their use in the policymaking process.

The indicators must be policy-relevant. Hence they must be constantly re-examined and refined. Depending on the shifts in the policy agenda, new information is needed all the time, for example, data on the contexts of education, on the various aspects of public and private financing, on staffing, decision-making and the curriculum, and, most crucially, on the outcomes of education. Data on outcomes are particularly difficult to obtain. In education one must distinguish between outcomes in at least three levels: students, systems, and labor markets. All three areas are problematic, but the most difficult is by far student performance.

For the indicators on student achievement, the OECD so far has had to rely on secondary data—that is, the data collected by other agencies and for other purposes. It is clear that the match between the needs of the OECD and those of major partners such as IEA is not always perfect. International surveys of student achievement are mammoth and costly undertakings. They place high demands on schools, teachers, and students. Understandably, therefore, they are conducted only infrequently. This poses problems, however, because the outcome data on which the set of OECD indicators crucially depends, tend to be outdated and are available for only a limited number of subject areas. The current procedures for collecting international data on student achievement are not sufficiently effective. Achieving an adequate model for the collection of data in this area is a stupendous challenge; it can be faced only through concerted effort.

For the indicators to be policy relevant they must be up-to-date. Statistics that are several years old can be of interest, but they do not tell much about the effectiveness of recent reforms and current policy. This puts both the OECD and the countries under pressure to reduce the time lag between data collection and reporting. The ideal would be to allow no more than two years of delay. *Education at a Glance II* has demonstrated that this can be done, but at a high price. Quite a few countries were unable to provide up-to-date information, especially about educational spending. More certainly needs to be done in this area as well.

The quality of the data used in the calculations is a key to success. The data set must meet certain minimum, but high, standards related to the conceptual design, implementation, processing, and analysis of the program. Although much progress has been made, the reliability and validity of the data on which the OECD depends are still unsatisfactory in some cases. Ensuring adequate comparability also remains an issue. No doubt, the overarching comparability problem relates to ISCED. Major shortcomings will inevitably remain until the current classification system is revised.

Finally, time series and detailed data analyses are needed in order to understand developments in education policy. The emphasis so far has been on improving the quality of the information. This reflects a choice, since trend data for policy analysis can be assembled only once stable and robust definitions are in place. But the lack of time series and the generally weak analytical orientation of much current work on education indicators are increasingly seen as a hinder to progress. What does this mean for the priorities of future work? There are two main implications. The first concerns the design of a future information management system, and the second has to do with the way the indicators are studied and used in policy analysis and even in research of a more fundamental nature.

The ultimate goal of the OECD-led effort is to build an innovative information system in education that will connect data needs and data sources at the national and international level. Such a system would serve multiple functions, ranging from the identification of data needs and data collection, to data development and analysis. An education information system has a variety of functions to perform; data collection is only one among them. While building an information system requires development in all areas, ideally one cannot move rapidly in all directions at the same time. So far, the priority has been in the areas of data needs identification and data collection. A future information management system in education should facilitate data development and data analysis in a multi-level framework, to a greater extent than is the case at present. The paragraphs below offer a perspective on what might be attempted.

Relationships Among Indicators

Although an indicator that is presented alone can be informative, value added would be achieved if more were known about the nature and magnitudes of the relationships among the various measures of inputs, processes, and outcomes used in the OECD framework of education indicators. That not much progress has been made in this important area is due partly to methodological reasons and partly to a lack of stable data sets. However, as the methodology evolves and more stable and robust data sets become available for secondary analysis, progress can be expected to occur in this area as well.

A simple correlational analysis of the OECD indicators can reveal interesting hypotheses for research and theory building. An example and some suggestions for further research are presented below. The example concerns the question whether factors such as the timing of entry into early childhood education, the timing of primary school entry, and the amount of financial resources allocated to primary education, are systematically associated with the cognitive performances of primary school children.

Table 3.4 presents evidence on crossnational variations in the timing of early childhood education, the primary-school entry age, and the amount of financial resources invested in primary schooling. Early childhood education can begin at two years of age in France, Germany, Norway, Spain, and the United Kingdom. In Austria, Ireland, Italy, Japan, Portugal and the United States, children must be at least three years old before they can be enrolled in early childhood education. In Australia, Canada, and the Netherlands, parents are expected to enroll their children at the age of four years, and in Denmark and Switzerland at five years. In 1991, only Finland and Sweden considered that pre-school education starts at age six.

The differences among countries in the earliest age at which children can participate in pre-school activities are reflected in different enrollment rates in early childhood education. Virtually all of the 3-year-olds in Belgium and France were enrolled in pre-schooling in the 1990/91 school year. This held true for the 4-year-olds in New Zealand and the Netherlands, and for 6-year-old children in Sweden. However, the differences in enrollment rates are less marked and less consequential than they might appear at first sight. Because countries use different definitions to draw a line between childcare and pre-schooling, comparisons of enrollment levels can be regarded as misleading, since much of what is considered childcare in one country is classified as early childhood education in another.

Table 3.4. Timing and duration of pre-schooling, and per-pupil public expenditure (in US$) in primary education converted using PPPs, 1991

	Earliest entry in pre-school (age)	Latest entry in pre-school (age)	Unit cost in primary education (US$)
Austria	3	7	3573
Belgium	2	6	2143
Canada	4	6	*
Denmark	5	7	4397
Finland	6	7	4060
France	2	6	2591
Germany	2	7	2609
Ireland	3	7	1542
Italy	3	6	*
Japan	3	6	3300
Netherlands	4	6	2791
New Zealand	2	5	*
Norway	2	7	3885
Portugal	3	6	2113
Spain	2	6	1861
Sweden	6	7	5470
Switzerland	5	7	5447
United Kingdom	2	5	2794
United States	3	6	5177

* Not provided.
Source: OECD (1993a, Tables P6 and P12).

Countries differ also with respect to the ages at which children are expected to enrol in the first grade of compulsory schooling, but the differences are much smaller than for pre-school education. Primary schooling normally starts at age 6 in the majority of the OECD countries. Children begin one year earlier in New Zealand and the United Kingdom, and one year later in Switzerland and the Nordic countries, where primary school begins at age 7.

In 1991, the typical OECD country spent about $3 340 per pupil (in equivalent US dollars and converted using PPP indices) at the primary level. This figure includes expenditures from public sources for pupils attending both public and private schools. However, this average level masks significant variations among countries: the cost per primary pupil varied from $5470 in Sweden to $1542 in Ireland. Thus, the outlays per primary pupil in Sweden, the highest spender, were more than three times as high as the level of expenditure in some other European countries. The wide variation in per-pupil expenditures reflects both differences in the physical quantities of the resources that countries provide to their primary schools—dissimilar pupil:teacher ratios, for example—and differences in the relative prices of resources, such as the level of teachers' salaries relative to the general price level in the same country.

The key question is whether the above factors are systematically associated with student performance in three major subjects: mathematics, science, and reading. The country mean scores derived from an international survey of the mathematics and science achievement of 13-year-old students, which was conducted as part of the International Assessment of Educational Progress (Lapointe, Mead, & Askew, 1992), and similar scores taken from an IEA study of the reading performance of 14-year-olds (Elley, 1992), can be employed to explore the hypothesis that the timing of school entry and the level of spending on primary education are associated with student performance.

Table 3.5 shows the overall country mean scores on the mathematics, science, and reading tests. The IEA reading test was administered in 18 OECD countries between October 1990 and April 1991, depending on the country concerned (Postlethwaite & Ross, 1992). Eleven OECD countries, or parts thereof, took part in the mathematics and science tests administered by the International Assessment of Educational Progress (IAEP:II) between September 1990 and March 1991 (Lapointe et al., 1992).

With respect to reading literacy, the country mean scores cluster into three loosely defined groups: students in Finland, France, Sweden, New Zealand, Switzerland and the United States generally outperform their peers in the other participating countries. Students in Spain and Belgium (French Community only) have significantly lower scores in reading than those in the other countries (OECD, 1993).

Multiple comparisons of the overall mathematics achievement of 13-year-olds show that students in the 15 participating cantons of Switzerland, as a group, clearly outperform their peers in the other countries. Students in Canada, France, Ireland, and England appear to have quite similar results, whereas students in Portugal, Spain (except Catalonia) and the United States have significantly lower scores on the mathematics test compared with the other countries (OECD, 1993).

The results in Table 3.5 furthermore suggest that the differences between the countries in science achievement are mostly small, and may be due to errors of sampling and measurement. However, as is the case with mathematics, the students in the 15 Swiss cantons achieved at a higher level on the science test than the students in the other countries.

Students in Ireland and Portugal, on the whole, appear to have significantly lower proficiency scores in science than students of the same age in the other participating countries (OECD, 1993).

Table 3.5. Country mean scores of 13- and 14-year-olds on international assessments of reading, mathematics, and science performance, 1991

	Student performance indicators		
	Reading (age 14)	Mathematics (age 13)	Science (age 13)
Belgium	446	—	—
Canada	494	620	688
Denmark	500	—	—
Finland	545	—	—
France	531	642	686
Germany	498	—	—
Ireland	484	605	633
Italy	488	640	699
Netherlands	486	—	—
New Zealand	528	—	—
Norway	489	—	—
Portugal	500	483	633
Spain	456	554	676
Sweden	529	—	—
Switzerland	515	708	737
United Kingdom	—	606	687
United States	514	553	670

Notes: — System scores are not available.
The standard errors of the estimates are given in OECD (1993a).
Source: OECD (1993a, Tables R1, R2 and R3).

It is clear from the previous discussion that the OECD countries apply different criteria in classifying programs as either childcare or pre-schooling. They also differ in how many children in an age group they enroll and how much they invest in primary schooling. Countries also tend to have different goals and objectives for pre-school activities, so that in defining the content and approach to education for different age groups, countries differ in the emphasis they place on the child's emotional and social development compared to cognitive development, the mastery of linguistic skills, and the ability of the child to generalize, to formulate ideas, and to solve problems. Despite these differences in goals and emphases, however, all countries consider the cognitive performance of school children as an important outcome indicator.

Table 3.6 presents a simple matrix of zero-order correlation coefficients calculated for the input and outcome indicators discussed above. It should be noted that these coefficients cannot be used for the making of causal inferences. One reason is that the data are measured at only one point in time. Moreover, very few cases undergird the coefficients, because the unit of analysis is the country level. Yet, despite these and other limitations, some interesting hypotheses for in-depth research can be derived from the estimates.

Table 3.6. Zero-order correlation coefficients between three predictor variables and three outcome variables, 1991

	1	2	3	4	5	6
1. Math	1.00					
2. Science	.80	1.00				
3. Reading	.32	.27	1.00			
4. Entry 0	.45	.44	.49	1.00		
5. Entry 1	.37	.05	.11	.49	1.00	
6. Costs 1	.43	.68	.60	.66	.39	1.00

Key to mnemonics:
1. Math: Mean score of 13-year-olds on a mathematics test.
2. Science: Mean score of 13-year-olds on a science test.
3. Reading: Mean score of 14-year-olds on a reading test.
4. Entry 0: Earliest age of entry into early childhood education.
5. Entry 1: Earliest age of entry into primary education.
6. Costs 1: Unit cost in primary education, in US dollars.

The estimates show that mathematics and science achievement correlate to a high degree (0.80). This could mean that countries whose 13-year-olds do well in mathematics are likely to do well also on the science test. The correlation between mathematics and science achievement, on the one hand, and reading performance on the other is significant (0.32 and 0.27), albeit less pronounced than between mathematics and science. A speculative interpretation of this result could be that it is due to the fact that different age groups are being compared, or that the number of countries for which complete data are available is too small to detect a stable association. Whatever the explanation, it would be of interest if the stability of student achievement scores in different subjects could be examined using high-quality data at the country level.

Another interesting—if not provocative—finding is that the earliest age at which children can enter pre-schooling appears to be consistently related to student performance in all three subjects: mathematics (0.45), science (0.44), and reading literacy (0.49). The size of these correlations and the consistency of the pattern are intriguing. However, these results appear to go against the findings of much previous research and therefore warrant to be studied more carefully.

Interesting also is the finding that the unit cost in primary education seems to be related to all three outcome variables: mathematics (0.43), science (0.68), and reading (0.60). These findings are of substantive interest, but they must be validated before any conclusions can be drawn. In a recent issue of *Educational Researcher,* Wainer (1993) reports a positive, albeit rather weak correlation between the states' scores on the NAEP assessment in mathematics and these states' per-pupil expenditures.

Conclusion

Since the late 1980s, important developments and variations in definitions, systems, and approaches to indicators have emerged. Recent work by international networks of countries and of individual countries building upon this work are highly encouraging. The quality and extent of information are increasing; and countries are adopting intentionally standard indicators and at the same time inventing and adapting other indicators for their own purposes. These innovations in turn may be adopted or adapted by other countries and other international agencies in the future. Other things being equal, the more points of valid comparison, the more useful the sets of indicators—although simplicity, parsimony, and concision can be virtues for some purposes. In these ways, countries and international agencies can continue to stimulate one another and develop more comprehensive, valid, and useful indicator sets.

References

Burstein, L., Oakes, J., & Guiton, G. (1992). Education indicators. *Encyclopedia of Educational Research* (sixth edition, pp. 409-418). New York: MacMillan Publishing Company.

Cohen, D.K., & Spillane, J.P. (1994). National education indicators and traditions of accountability. In A.C. Tuijnman and N. Bottani (Eds.), *Making education count: Developing and using international indicators.* Paris: OECD, Center for Educational Research and Innovation.

Darling-Hammond, L. (1994). Policy uses and indicators. In A.C. Tuijnman and N. Bottani (Eds.), *Making education count: Developing and using international indicators.* Paris: OECD, Center for Educational Research and Innovation.

De Groof, J., & Van Haver, T. (Eds.) (1993). *De school op rapport. Het Vlaams onderwijs in internationale context.* Brussels: Ministry of the Flemish Community, Department of Education. Kappellen: Uitgeverij Pelckmans.

Denmark, Ministry of Education and Research (1993). *Facts and figures. Education indicators. Denmark.* Copenhagen: Author.

Elley, W.B. (1992). *How in the world do students read?* The Hague: The International Association for the Evaluation of Educational Achievement.

France, Ministry of National Education (1992a). *L'état de l'école. Janvier 1992.* Paris: Ministère de l'Education Nationale, Direction de l'Evaluation et de la Prospective.

France, Ministry of National Education (1992b). *L'état de l'école. Octobre 1992.* Paris: Ministère de l'Education Nationale, Direction de l'Evaluation et de la Prospective.

France, Ministry of National Education (1993a). *L'état de l'école. 1993.* Paris: Ministère de l'Education Nationale, Direction de l'Evaluation et de la Prospective.

France, Ministry of National Education (1993b). *La géographie de l'école.* Paris: Ministère de l'Education Nationale, Direction de l'Evaluation et de la Prospective.

van Herpen, M. (1992). Conceptual models in use for education statistics. In OECD, *The OECD international education indicators: A framework for analysis* (pp. 25-51). Paris: OECD, Center for Educational Research and Innovation.

Johnstone, J.N. (1981). *Indicators of education systems.* London: Kogan Page.

Lapointe, A.E., Mead, N.A., & Askew, J.M. (1992). *Learning mathematics.* Princeton, New Jersey: The International Assessment of Educational Progress, Educational Testing Service.

Nordic Council (1991). *Educational indicators in the Nordic countries. Describing educational status and student flows.* Copenhagen: Nordic Statistical Secretariat, and Stockholm: Norstedts Tryckeri.

Nuttall, D. (1992). The functions and limitations of international education indicators. In OECD, *The OECD international education indicators: A framework for analysis* (pp. 13-21). Paris: OECD, Center for Educational Research and Innovation.

OECD (1992a). *Education at a glance I.* Paris: OECD, Center for Educational Research and Innovation.

OECD (1992b). *The OECD international education indicators: A framework for analysis.* Paris: OECD, Center for Educational Research and Innovation.

OECD (1992c). The Ministers' communiqué. In *High-quality education and training for all* (pp. 31-36). Paris: OECD.

OECD (1993a). *Education at a glance II.* Paris: OECD, Center for Educational Research and Innovation.

OECD (1993b). *Education at a glance I in the press.* Paris, OECD, Press Division.

OECD (1994). *Education at a glance II in the press.* Paris, OECD, Press Division.

Pelgrum, H., and Plomp, T. (Eds.) (1993). *The IEA study of computers in education: Implementation of an innovation in 21 education systems.* Oxford: Pergamon Press.

Postlethwaite, T.N., & Ross, K.N. (1992). *Effective schools in reading: Implications for educational planners.* The Hague: The International Association for the Evaluation of Educational Achievement.

Switzerland, Federal Office of Statistics (1993). *Les indicateurs de l'enseignement en Suisse.* Bern: Federal Office of Statistics.

UNESCO (1992). *World education indicators.* Paris: United Nations Educational, Scientific, and Cultural Organization.

UNESCO (1993). *World report on education.* Paris: United Nations Educational, Scientific, and Cultural Organization.

US, National Center for Education Statistics (1992). *The condition of education 1992.* Washington, DC: Office of Educational Research and Improvement, US Department of Education.

US, National Center for Education Statistics (1993). *Education in states and nations: Indicators comparing US states with the OECD countries in 1988.* Washington, DC: Office of Educational Research and Improvement, US Department of Education.

US, National Education Goals Panel (1991). *Building a nation of learners: The national education goals report, 1991.* Washington, DC: US Government Printing Office.

US, National Education Goals Panel (1992). *The national education goals report, 1992.* Washington, DC: US Government Printing Office.

US, National Education Goals Panel (1993). *The national education goals report. Volume one: The national report, 1993.* Washington, DC: US Government Printing Office.

Wainer, H. (1993). Does spending money on education help? A reaction to the Heritage Foundation and the Wall Street Journal. *Educational Researcher* 22(9), 22-24.

Chapter 4

Toward an Empirical Taxonomy of World Education Systems

HERBERT J. WALBERG, GUOXIONG ZHANG and
VERNON C. DANIEL

University of Illinois at Chicago, USA

This study first sets the stage for a taxonomy of national education systems by briefly discussing important milestones in the history of natural science taxonomies and analyzing five previous educational taxonomies that make use of a variety of educational, cultural, political, and social concepts. A taxonomic analysis is made of a UNESCO data base of 139 world education indicators for 164 countries. The indicators include literacy rates, enrollment at several levels of education, pupil:teacher ratios, and other measures. The analysis shows that the national systems may be parsimoniously and discretely clustered into three major groups: 79 functional democracies mostly in Asia, Europe, Oceania, and North and South America; 36 post-industrial democracies in Northern Europe and North America; and four clusters of 49 underdeveloped countries largely in Africa.

By definition, the field of comparative education concerns national comparisons. In accomplishing this end, case studies of single education systems can be useful because they identify characteristics of systems that may set them apart from others. Often reported by foreigners, such studies commonly make implicit comparisons between the observer's (typically Western) country and the observed country. Still, it takes explicit observations of two or more countries to make replicable comparisons.

Observing a single educational characteristic of two countries yields one comparison. Four countries yield six possible comparisons, and the number of comparisons escalates rapidly with additional numbers of countries. The 24 Member countries of the Organization for Economic Cooperation and Development (OECD), for example, may be compared in 276 ways. The 167 countries in UNESCO's (1992) data bank of world education indicators allows 13,861 comparisons. Such comparisons, of course, rise multiplicatively with the number of education system features compared. Since UNESCO provides data on 139 educational characteristics, the total number of possible comparisons of the countries is more than 1.9 million.

How many comparisons can be assimilated by those who seek an understanding of the patterns of comparisons? Writers on psychology such as Nobel laureate Herbert Simon (1983) believe that humans can simultaneously consider only a few items of information, perhaps fewer than four (depending partly on the "chunking" of information). For this reason, humans employ categorical ideas to think efficiently about what would be overly complex, possibly incomprehensible matters of the infinitely variegated real world. In formal, especially scientific theory, taxonomies—that is, explicitly-defined categorical systems—enable the rigorous sorting of hosts of entities into a limited number of mutually exclusive categories with common elements. To consider the potential contribution of taxonomies to educational thought, a few highlights of the history of taxonomic insights in the natural sciences may be instructive.

Definitions and Taxonomic Insights

Aristotle's definition of humans as "featherless bipeds" set forth a parsimonious scheme placing humans within an animal category but setting them apart from all two-legged others (insofar as could be ascertained in his time). His taxonomies of plants, animals, and the constitutions of Greek city-states were major accomplishments of classical times. His categorical themes dominated Western thought for more than a millennium.

In the modern era, the Swedish botanist Carolus Linnaeus (1707-78) challenged Aristotle's taxonomies. Founder of the binomial nomenclature, he established an enduring means of classifying plants, animals, and diseases. Charles Darwin (1809-82), the English naturalist, showed how classifications may differ across time and place and established the theory of organic evolution. As naturalist aboard H.M.S. Beagle during its world voyage (1831-36), he began assimilating vast zoological data about corral reefs, finches, and other animal life. His reflections on such data led to his simultaneous discovery and publication with Alfred Wallace in 1858 of the theory of natural selection and the origin of species. Finally, of the founding natural science taxonomists, the Russian Dimitri Mendeleev (1834-1907) invented the periodic table, a system of periodic laws and classification of the elements that allowed chemists to predict properties of unknown elements (see the Concise Columbia Encyclopaedia, 1991, and more detailed reference works for treatments of such classic contributions to the history of taxonomic thought).

Perhaps the best known taxonomy in education is Bloom's (1956) classification of cognitive objectives ranging from knowledge and comprehension to synthesis and evaluation. His taxonomy encouraged test makers to compose more items to measure "the higher processes" and educators to allocate more time and energies to activities that improve them. Taxonomies of education systems are discussed below (for recent essays on Bloom's and other learning taxonomies, see Anderson & Sosniak, 1994).

The insights of founding taxonomists endured because they enabled scientists and others to think clearly, concisely, and systematically about the enormous amounts of chaotic data that fill the natural world. With a few analytic distinctions, a taxonomy can capture the salient characteristics that establish commonalities and differences among entities. Since common features are often correlated, it is possible to classify entities on the basis of only a few characteristics, which allows economy of thought and effort. On this basis, an expert (and sometimes even a quick, well-tutored novice) can readily distinguish cases, attain understandings, make predictions, and take useful action. Such actions can even be routinized: As Alfred North Whitehead (1861-1947) declared, "Civilization advances by extending the number of important operations which we can perform without thinking about them" (quoted in DeWolf, 1980).

Still, taxonomies can be misleading since they risk oversimplification and can lead to flawed thinking. A characteristic thought to be distinguishing can be spurious; and not all units in a category are necessarily identical. The term "stereotype" conveys the danger of false, pejorative classification—a spurious notion that if entities possess a particular neutral character, they must also have an undesirable trait, as in the case of ethnic stereotypes.

In a statistical sense, a taxonomy can be thought useful if it can parsimoniously explain considerable variance among entities with only a few categories and distinguishing characteristics. As in analysis of variance (ANOVA), for example, the ratio of among-classification variation to within-classification variation is a key and objective indicator of taxonomic worth. That is, well-classified entities within any category should be similar to each other and different from those in other categories (cf Dixon & Massey, 1975, and other works on ANOVA).

Education Systems

Education systems may be thought of as elements in a changing system of interacting relationships that involve the economic, psychological, social and political dimensions of societies (Fägerlind & Saha, 1983; Hallak, 1990). Historical attributes and tensions in the society produce education systems with both unique and common features in the various countries and regions of the world (Inkeles & Serowy, 1983).

No well-defined nomenclature describes education systems nor their links to social and economic development (McGinn, 1978). Nor do systematic sets of propositions describe the processes of change in the formal and nonformal parts of education systems (Adams, 1981). Therefore, empirical studies seem in order that would provide generalizable explanations for differences among education systems. These studies might increase our understanding and provide empirical evidence that supports education and economic development strategies (Hallak, 1990). According to Walberg (1989), "how changes in education might affect our society and economy is worth thinking about" (p. 106).

Researchers have tried to explain education as a phenomenon occurring in the economic, cultural and political contexts of various countries (Ginsburg, Cooper, Raghu, & Zegarra, 1990). Several authors—whose work is reviewed below—use taxonomic concepts to order and classify education systems (Paulson, 1982). Taxonomies can have descriptive utility because the creation of categories involves purposive abstraction of empirical referents (McKinley, 1966). Taxonomies can organize diverse measurements of the cultural characteristics of modern nations (Cattell & Brennan, 1984), governmental action (O'Here, 1989), and Third World conflicts (Djalili, 1991). In educational research, taxonomies are heuristic devices that can inform educational planning (Paulson, 1982; Fägerlind & Saha, 1983).

System Taxonomies

While national education systems may share similar characteristics (Inkeles & Sirowy, 1984), taxonomies could make clear the way in which these systems are alike and highlight the differences among them. Taxonomies might increase comparative researchers' understanding of education systems in the regions and countries of the world. When more than one school system is analyzed, researchers can use the variance that naturally exists among them to construct multiple types (Loftland & Loftland, 1984). Constructed categories with central tendencies give types their meaning or aggregate characteristics (Paulson, 1982). They can serve as a method for abstracting phenomena (McKinley, 1966) and a means with which to achieve a scientific understanding of natural and social phenomena (Lazarsfeld, 1967). Further, typologies can be used as proto-theory to make development work theoretically grounded and rational (Paulson, 1982).

The dimensions of education systems that define the various types should be clearly and explicitly stated (McKinley, 1966); each discrete type should be mutually exclusive; and the taxonomy must be comprehensive enough to classify all or almost all relevant cases (Tiryakian, 1974; Loftland & Loftland, 1984). In addition, taxonomies should have heuristic significance that facilitates the understanding of relationships between types, and a parsimonious character that limits the number of significant types required to organize the observed data (Paulson, 1982).

Previous System Taxonomies

Comparative education is a multidisciplinary field that provides a structured approach to the study of educational problems and issues between and among countries (Spaulding, 1989, p. 1). Researchers in the area provide an international perspective on the provision of educational services in individual countries (Altbach, 1991). Many use implicit taxonomies to clarify differences between systems in developed and developing societies (Weiler, 1979). Harbison and Myers (1964), for example, employed 14 social, economic and educational indicators to develop a composite index of human resource development. This index is shown in Table 4.1.

Table 4.1. Economic taxonomy of human resource development

Level	Number of countries	Human development level
I	17	Underdeveloped
II	21	Partially developed
III	21	Semi-advanced
IV	16	Advanced

Source: Harbison and Myers (1964).

The composite index is based on the weighted characteristics of enrollment at the secondary and tertiary levels of education systems. In formulating this index, Harbison and Myers broadened the concept of education to include human resource development and argued that it influences economic growth. Their taxonomy is based on empirical data from both developing and developed countries.

Beeby (1966) proposed an evolutionary taxonomy to analyze the character of change in education systems. He argued that school systems pass through predetermined and qualitative stages in a hierarchy of four discrete types—dame schools, formalism, transition, and meaning—and that the general education level of teachers and their professional training determine the ability of education systems to move up in the hierarchy. Beeby's taxonomy is reproduced in Table 4.2.

Focusing on classroom transactions, Thomas (1968) developed a taxonomy based on the mode of instruction that characterises education systems. As can be seen from Table 4.3, he proposed four discrete instructional types—memorizing, training, intellect development, and problem solving—and employed seven identifying principles to differentiate among them: sources of the curriculum, character of goals, curriculum content, learning activities, teaching methods, tests of success, and anticipated outcomes.

Table 4.2. Taxonomy of educational evolution

Stage	Teacher characteristics	School system characteristics
Dame school	Limited education, untrained	Unorganized, meaningless symbols, very narrow subject content, very low standards
Formalism	Limited education, trained	Highly organized, symbols with limited meaning, rigid syllabus, memorizing stressed
Transition	Better-educated, trained	Highly organized, more emphasis on meaning, less restrictive syllabi and textbooks
Meaning	Well-educated, well-trained	Meaning and understanding stressed, wider curriculum, variety of content and methods, problem solving and creativity stressed

Source: Beeby (1966).

Adams and Farrell (1969) featured the distinction between traditional and modern educational strategies to develop a taxonomy of educational differentiation. Table 4.4 presents this taxonomy. They hypothesized four levels of educational differentiation: socialization through the family, socialization differentiated out of the family and provided for all adolescent members of the society, control of education by the elites in the society, and complex relationships between education and other social structures.

Table 4.5 presents a taxonomy of education systems based on national economic types and socioeconomic policy orientations proposed by McGinn and Snodgrass (1979). They proposed six types according to the relative priority countries place on the production and distribution of raw materials extraction, agricultural and intermediate processing, and import substitution.

Table 4.3. Curriculum-based taxonomy

School system type	Character of goals	Development breadth
Memorizing	Transmitting the values of the past	Efficient memorizing of the best in the cultural heritage
Training	Enabling children to learn to do better what they will have to do	Developing a specific set of skills accompanied by appropriate attitudes
Developing	Providing for intrinsic, preparatory and individualized development of students	The development of basic equipment and subject familiarity needed by prospective specialist
Problem solving	Transmitting the core values of the society and nature of the problem-solving process	The discovery of cultivated enjoyments, developing democratic self-concepts and socialized personalities

Source: Thomas (1968).

Table 4.4. Structural differentiation taxonomy

Level of educational differentiation	Structural differentiation of socialization system	Education system function
I	Socialization through the family	Provide economic skills and introduction to homogeneous social life
II	Socialization process differentiates out of the family	Strong emphasis on the metaphysical and conduct areas
III	Education linked to small groups who wielded political, economic or religious power	Emphasis on social, political and economic hegemony
IV	Education linked to industrialization and ever increasing social differentiation	Provide for the division of labor and role specialization

Source: Adams and Farrell (1969).

Table 4.5. Structural-functional taxonomy

Type of education system	Economic policy	Policy orientation	Production factor
I	Export promotion	Emphasis on production	Raw materials extraction
II	Export promotion	Emphasis on distribution	Raw materials extraction
III	Export promotion	Emphasis on production	Agriculture and intermediate processing
IV	Export promotion	Emphasis on distribution	Agriculture and intermediate processing
V	Import substitution	Emphasis on production	Imported goods and services
VI	Import substitution	Emphasis on distribution	Imported goods and services

Source: McGinn and Snodgrass (1979).

Oddly, the foregoing systems taxonomists seemed to have paid little attention to one another' work. Therefore, their conceptual frameworks of the studies cannot be easily linked together to develop a consensually-based taxonomy, although their essentials can be concisely stated. Table 4.6 summarizes the main characteristics of the five taxonomies of education systems mentioned previously. Harbison and Myers (1964) associated the evolution of school systems with economic growth; Thomas (1968) proposed a cross-cultural typology related to modes of instruction; Beeby (1966) suggested that education systems evolve through predetermined stages; Adams and Farrell (1969) linked education systems to social differentiation; McGinn and Snodgrass used a structural-functional framework to classify education systems.

The typologies are somewhat incomparable because they are based on a different social, economic and educational constructs. Some are framed as hierarchical or developmental paradigms; other are not. With the exception of Harbison and Meyers, the taxonomies have apparently not been subjected to empirical falsification tests, presumably the *sine qua non* of science. Like the insights of "armchair psychology," they yield useful insights and are probably based on direct experience and extensive scholarship. But they do not appear to have been explicitly tested with wide-scale system data.

In contrast, the present research takes as its starting point empirical observations thought important enough by ministries of education and international development agencies to collect on a systematic and world-wide basis. It applies objective mathematical procedures to determine how the countries cluster together strictly on the basis of their characteristics. In addition to the usual purposes of taxonomies discussed above, the exercise is intended to shed light on the possible validity of previous taxonomies, although they cannot be rigorously tested because several require data that is yet unavailable for a large collection of countries.

Method

This study compares countries in order to gain insight into the states of education around the world, and to create a taxonomy of global education. The data were obtained from UNESCO's Division of Statistics which assembled figures from its existing statistical databases to create a set of world education indicators (UNESCO, 1992). This indicator system provided not only education indicators per se, but also a framework which allows the interpretation of one figure (say, enrollment) in the context of the demography and investment in education of a particular country.

Table 4.6. Characteristics of the taxonomies

Taxonomists	Conceptual framework	Organizing constructs
Harbison and Myers (1964)	Evolution based on economic growth	Social, education and economic indicators
Beeby (1966)	Evolution through predetermined stages	Teachers' level of education and training
Thomas (1968)	Modes of instruction	Education goals, the curriculum, teaching and learning activities, tests of success
Adams and Farrell (1969)	Differentiated education in societies	The social function of education
McGinn and Snodgrass (1979)	National economic type and socio-economic policy orientation	Policy and orientation that controls the factors of production

Among the key indicators in the data set are literacy rates, enrollment at different levels of education, and information about the teaching force. There also is contextual information about the characteristics of the population of each country along with its age distribution, and information about public expenditure on education. Most of the indicators are given for both 1980 and 1988. In some cases, more recent years are also available. Some breakdowns, moreover, are given by gender. Appendix 4.1 provides a complete list as well as the definitions of all variables used for the data analysis reported in this chapter.

The data consist of 139 indicator variables for 167 countries. A few missing values of some of the variables were substituted with the mean values (of all countries with non-missing data) so as to keep the sample as large as possible. Logarithmic (base 10) transformation was used on such variables as GNP per capita to avoid high positive skewness. In all cases, the most recent information was selected, but for the several variables on which observations for separate years were available, the percentage gain value between two years was calculated. To facilitate comparisons of variables with unlike scales, the data were standardized to *T* scores with mean values of 50 and standard deviations of 10 units. In discussing the results, both the raw averages and the standardized averages are reported.

Cluster procedures were used for grouping or classifying countries based on the UNESCO data set. The FASTCLUS program in SAS (1985, pp. 377-400) was chosen because it is a standard procedure especially suited for large data sets. FASTCLUS divides all observations into mutually exclusive sets on the basis of computed Euclidean distances among them on the criterion variables.

A computer was programmed to yield up to ten clusters which would include as many or more classifications as previous education taxonomies. The program yielded nine clusters. However, two African countries, Burundi and Djibouti, and one Asian country, Oman, each turned out to form a single unique country cluster. For this reason, they were not included in the final data analysis. The remaining six clusters included from 5 to 79 countries.

Results

Table 4.7 shows the six clusters revealed by the analysis and the distribution of countries in clusters in accordance with UNESCO's classification of world areas. The first cluster, termed *Functional Democracies*, groups together almost half (79 of 164) of the countries in the world. Asia had 32 percent (63 percent of all) of its countries in this cluster; North America had 25 percent (80 percent of all); Africa, 18 percent (28 percent of all); South America, 11 percent (75 percent of all); Europe, 9 percent (23 percent of all); and the Oceanic Region 5 percent (57 percent of all).

Table 4.8 lists the countries in the cluster and shows their distinguishing features. The functional democracies are characterized by high values on primary education indicators: total intake rate and gross enrollment ratio, coefficient of completion efficiency, and percentage of female teachers. These countries also had high percentages of females in such fields of tertiary study as law and social sciences. These countries also had low illiteracy rates for the group aged 15 to 19 and total population, percentage gains of male over female cohorts reaching grade four, and population dependency ratio gains for ages 0 through 14.

The second largest cluster, termed *Post-Industrial Democracies* in Tables 4.7 and 4.8, contains many of the world's affluent and educationally developed countries. This cluster has 77 percent (23 of 30) of the European countries such as Sweden and five Asian countries including Australia, Japan, and New Zealand. In North America, the cluster includes the United States and Canada.

These post-industrial countries are high in GNP per capita and enrollments in secondary and tertiary education and for the group aged 4 to 23 years. As might be expected, they are also high in per-capita paper consumption. These countries are low in the average annual population growth rate, population dependency ratio, elementary and secondary pupil:teacher ratios, and percentages of elementary-school grade-repeaters.

Table 4.7. Number of countries by cluster and world region

Cluster	Asia	Europa	Africa	North America	Oceania	South America	Row total
Functional democracies	**25**	7	**14**	**20**	**4**	**9**	79
Post-industrial democracies	5	**23**	0	**4**	2	2	36
Underdeveloped (a)	5	0	**21**	0	0	0	26
Underdeveloped (b)	1	0	**9**	0	1	0	11
Underdeveloped (c)	**3**	0	**3**	0	0	1	7
Underdeveloped (d)	1	0	**3**	1	0	0	5
Column total	40	30	50	25	7	12	164

Note: Cell entries for continents with relatively large numbers of countries for each cluster are in bold typeface.

The remaining four clusters, labelled *Underdeveloped a-d* in Tables 4.7 and 4.8 were constituted mostly of African countries. The biggest group, designated *Underdeveloped a* included 21 African countries such as Chad and Ghana, and five Asia countries such as Afghanistan and Morocco. They are distinctive in having high rates of illiteracy, grade repeating, gains in dependency, high pupil:teacher ratios in primary schools, and low rates of primary, secondary, and tertiary enrollments and ratios of primary female teachers.

The second group of underdeveloped countries, designated *Underdeveloped b* in Tables 4.7 and 4.8 included only nine countries, for example, Bangladesh and Burkina Faso. They are characterized by high pupil:teacher ratios and illiteracy. They gained considerably in paper consumption (kgs. per 1000 inhabitants) and television exposure in recent years, but had poor rates of newspaper circulation. These countries had low ratios of female primary teachers and female enrollments in tertiary education.

Table 4.8(a). Distinguishing features of clusters: Functional democracies

Standard mean	Raw mean	Top and bottom five variables
54	96.7	1st level apparent intake rate, total 1988
54	90.8	1st level gross enrollment ratio, total 1988
53	.8	1st level, coefficient of efficiency, 1988
53	55.9	% 1st level female teachers, 1988
53	43.2	% females in 3rd level law and social sciences
47	16.3	Illiteracy rate, ages 15–19, 1990
47	.3	% 1987 cohort reaching grade 4, male-female
47	3.7	Total population, 1990
47	36.9	Illiteracy rate, total population, 1990
46	-3.5	Dependency ratio gain, ages 0–14, 1990–1980

Note: Countries: Albania (EP), Algeria (AF), Angola (AF), Antigua & Barbuda (NA), Bahamas (NA), Bahrain (AS), Barbados (NA), Belize (NA), Bolivia (SA), Brazil (SA), British Virgin Islands (NA), Cambodia (AS), Cape Verde (AF), China (AS), Colombia (SA), Costa Rica (NA), Cyprus (AS), Former Dem. Yemen (AS), Dominica (NA), Dominican Republic (NA), Ecuador (SA), Egypt (AF), El Salvador (NA), Equatorial Guinea (AF), Fiji (OC), Palestine, Gaza Strip (AS), Grenada (NA), Guatemala (NA), Guyana (SA), Honduras (NA), Hong Kong (AS), Iceland (EP), India (AS), Indonesia (AS), Iran (AS), Jamaica (NA), Kiribati (OC), Korea, Dem. People's Republic (AS), Lebanon (AS), Lesotho (AF), Libyan Arab Jamahiriya (AF), Luxembourg (EP), Malaysia (AS), Maldives (AS), Mauritius (AF), Mexico (NA), Monaco (EP), Mongolia (AS), Myanmar (AS), Namibia (AF), Netherlands Antilles (NA), Nicaragua (NA), Paraguay (SA), Peru (SA), Philippines (AS), Portugal (EP), Romania (EP), Samoa (OC), San Marino (EP), Sao Tome and Principe (AF), Saudi Arabia (AS), Seychelles (AF), Singapore (AS), South Africa (AF), Sri Lanka (AS), St. Kitts Nevis (NA), St. Lucia (NA), St. Vincent & Grenadines (NA), Suriname (SA), Swaziland (AF), Syrian Arab Republic (AS), Thailand (AS), Tonga (OC), Trinidad & Tobago (NA), Tunisia (AF), Turkey (AS), Venezuela (SA), Vietnam (AS), Palestine, West Bank (AS).
Legend: AS=Asia; EP=Europe; AF=Africa; NA=North America; OC=Oceania; SA=South America.

The seven countries counted in *Underdeveloped c* included only Chile in South America; Iraq, Jordan, and Qatar in Asia; and Sudan, Zaire, and Zambia in Africa. These countries gained in population, per capita newspaper consumption, percentage of private secondary school enrollments, and in female secondary teachers. They were low, however, in per capita growth rates in per capita GNP and public educational expenditure.

Table 4.8(b). Distinguishing features of clusters: Post-industrial democracies

Standard mean	Raw mean	Top and bottom five variables
63	13.7	3rd level, gross enrollment ratio, total 1988
63	51.7	2nd level, gross enrollment ratio, total 1988
62	3.2	GNP per capita, in US$, 1989
62	52.2	Gross enrollment ratio, ages 4–23, 1988
61	3.3	Paper consumption, 1988
44	12.7	1st level education, % repeaters, 1988
43	19.4	2nd level pupil:teacher ratio, 1988
41	30.1	1st level pupil:teacher ratio, 1988
40	2.0	Average annual population growth rate, 1980–90
37	64.1	Population dependency ratio, ages 0–14, 1990

Note: Countries: Argentina (SA), Australia (OC), Austria (EP), Belgium (EP), Bulgaria (EP), Canada (NA), Cuba (NA), Czechoslovakia (EP), Denmark (EP), Finland (EP), France (EP), Former Germany, Democratic Republic (EP), Former Germany, Federal Republic (EP), Greece (EP), Hungary (EP), Ireland (EP), Israel (AS), Italy (EP), Japan (AS), Korea, Republic of (AS), Kuwait (AS), Malta (EP), Netherlands (EP), New Zealand (OC), Norway (EP), Panama (NA), Poland (EP), Spain (EP), Sweden (EP), Switzerland (EP), United Kingdom (EP), United Arab Emirates (AS), Uruguay (SA), United States (NA), USSR (EP), Former Yugoslavia (EP).
Legend: see Table 4.8(a).

The fourth cluster of countries—*Underdeveloped d*—includes Botswana, Haiti, Uganda, (former) Yemen, and Zimbabwe. These countries made on average relatively high gains in efficiency in educational completion rates and gross enrollment gains. They had relatively few secondary female teachers and students in the sciences, engineering, and agriculture. They were also low in urban population and in paper and radio consumption.

Conclusion

The present study suggests that it is possible to usefully classify 164 world education systems into three broad types:

First, about half of the 164 countries may be characterized as *Functional Democracies*. They are found in all major parts of the world. Their economies include agriculture, industry, mining, and services. Their education systems appear extensive, and they are relatively egalitarian with respect to males and females.

Table 4.8(c). Distinguishing features of clusters: Underdeveloped A

Standard mean	Raw mean	Top and bottom five variables
61	12.7	Percentage of repeaters, 1988
60	-3.5	Dependency gain ratio, ages 0–14, 1990–80
60	64.1	Dependency ratio, ages 0–14, 1990
59	36.9	Illiteracy rate, total population, 1990
58	30.1	1st level pupil:teacher ratio, 1988
39	90.8	1st level gross enrollment ratio, total 1988
38	55.9	% 1st level female teachers, 1988
38	51.7	2nd level gross enrollment ratio, total 1988
38	2.8	Number of 3rd level students, 1988
38	52.2	Gross enrollment ratio, ages 4–23, 1988

Note: Countries: Afghanistan (AS), Bhutan (AS), Cameroon (AF), Chad (AF), Comoros (AF), Cote d'Ivoire (AF), Ethiopia (AF), Gabon (AF), Gambia (AF), Ghana (AF), Guinea-Bissau (AF), Kenya (AF), Lao People's Democratic Republic (AS), Liberia (AF), Madagascar (AF), Malawi (AF), Mali (AF), Morocco (AF), Mozambique (AF), Nepal (AS), Nigeria (AF), Pakistan (AS), Sierra Leone (AF), Somalia (AF), Togo (AF), Tanzania (AF).

Table 4.8(d). Distinguishing features of clusters: Underdeveloped B

Standard mean	Raw mean	Top and bottom five variables
67	30.1	1st level pupil:teacher ratio, 1988
65	36.9	Illiteracy rate, ages 15 and over, total 1990
64	.2	Paper consumption gain, 1980–1988
63	1.4	Television gain, 1980–1988
62	19.4	2nd level pupil:teacher ratio, 1988
35	55.9	% 1st level female teachers, 1988
34	3.3	Daily newspaper (copies/1000), 1988
33	43.2	% females in 3rd level law and social sciences
30	49.1	% females in 3rd level humanities
30	52.2	% females in 3rd level education

Note: Countries: Bangladesh (AS), Benin (AF), Burkina Faso (AF), Central African Republic (AF), Congo (AF), Guinea (AF), Mauritania (AF), Niger (AF), Papua New Guinea (OC), Rwanda (AF), Senegal (AF).
Legend: see Table 4.8(a).

Table 4.8(e). Distiguishing features of clusters: Underdeveloped C

Standard mean	Raw mean	Top and bottom five variables
63	.2	Daily newspaper gain (copies/1,000), 1988–80
62	61.9	2nd level enrollment gain
61	2.0	Average annual population growth rate, 1990–80
61	.3	% of 1987 cohort reaching grade 4, male-female
61	1.7	% gain 2nd level female teachers, 1988–80
40	2.8	Public education spending growth rate, 1980–88
40	.4	Average annual GNP per capita growth rate, 1980–89
36	-2.0	Pre-primary pupil:teacher ratio gain, 1988–80
33	-6.0	Teachers as % non-agriculture, male-female, 1988
32	-.9	2nd level net enrollment gain, ratio male–female, 1988–80

Note: Countries: Chile (SA), Iraq (AS), Jordan (AS), Qatar (AS), Sudan (AF), Zaire (AF), Zambia (AF).
Legend: see Table 4.8(a).

Table 4.8(f). Distinguishing features of clusters: Underdeveloped D

Standard mean	Raw mean	Top and bottom five variables
75	.0	1st level, coefficient of efficiency, gain 1988–80
72	.3	1st level gross enrollment ratio, gain 1988–80
68	3.5	Gross enrollment gain ratio age 4–23 1990–80
67	6.6	2nd level gross enrollment ratio, gain 1988–80
65	.7	2nd level private enrollment gain as % total 1988–80
40	39.1	% 2nd level female teachers, 1988
38	51.4	Percentage urban populations, 1990
38	3.3	Paper consumption, 1988
37	2.4	Radios, 1988
34	29.3	% students in natural sciences and engineering

Note: Countries: Botswana (AF), Haiti (NA), Uganda (AF), Former Yemen (AS), Zimbabwe (AF).
Legend: see Table 4.8(a).

Second, about a fifth of the world's countries are affluent *Post-industrial Democracies* concentrated in Asia, Europe, and North America. While they have employment in agriculture, mining, and industry, their economies are increasingly concentrated on services, particularly information services. Correspondingly their education systems are even more extensive and efficient with respect to completion rates than functional democracies.

Third, concentrated in Africa, the remaining 30 percent of counties are relatively "underdeveloped" both with respect to their economies—many of which are "pre-industrial"—and their education systems. The educational entry and completion rates are low, and females are extensively underserved. Among the four clusters of underdeveloped countries, however, some are making good educational and economic progress.

The five previously discussed taxonomies differ from one another and from the results obtained in this study. Nonetheless, they offer useful confirmatory perspectives on the findings. Harbison and Myers (1964) and McGinn and Snodgrass (1979) stress the interdependence of economies and education systems. Adams and Farrell (1969) emphasize family socialization leading to industrialization and the division of labor markets. The clusters identified in the present study provide support for such interdependence and evolutionary processes. The present study parsimoniously yields fewer clusters than these economic taxonomies, but a larger set, if that is desired, could be identified by separate cluster analysis of the sets of functional or post-industrial democracies.

In their taxonomies, Beeby (1966) and Thomas (1968) identify teacher training and classroom instruction as key differentiating features of national education systems. The present "macro" indicators cannot discriminate at the implied level of schools and teacher preparation institutions. It is not difficult, nonetheless, to imagine that the more sophisticated preparation and instruction called for in Beeby's and Thomas's taxonomies would be increasingly demanded in the post-industrial democracies. Detailed information on a set of countries might usefully be employed to test this hypothesis.

References

Adams, D. (1981). *Educational systems, educational planning, and educational indicators.* Pittsburgh, PA: International and Development Program, University of Pittsburgh.

Altbach, P.G. (1991). Trends in comparative education. *Comparative Education Review* 35(3).

Anderson, L.W., & Sosniak, L.A. (1994). *Bloom's taxonomy of educational objectives.* Chicago, IL.: The University of Chicago Press.

Beeby, C.E. (1966). *The quality of education in developing countries.* Cambridge, MA: Harvard University Press.

Bloom, B.S. (Ed.) (1956). *Taxonomy of educational objectives: Cognitive domain.* New York: Longmans Green.

Cattell, R.B., & Brennan, J. (1984). The cultural types of modern nations, by two qualitative classification methods. *Sociology and Social Research.* 68(2), 208-235.

DeWolf, A.S. (1980). *Bartlett's familiar quotations.* Boston, MA: Little, Brown, and Company, Inc.

Dixon, W.J., & Massey, F.J. (1975). *Introduction to statistical analysis.* New York: McGraw-Hill.

Djalili, M.-R. (1991). Analysis of third world conflicts: Outline of a typology. *International Social Science Journal 43*, 163-171.

Fägerlind, I., & Saha, L.J. (1983). *Education and national development: A comparative perspective.* Oxford: Pergamon Press.

Ginsburg, M.B., Cooper, S., Raghu, R., & Zegarra, F. (1990). National and world system explanations of educational reform. *Comparative Education Review* 34(4), 474-502.

Hallak, J. (1990). *Investing in the future: Setting educational priorities in the developing world.* Paris: UNESCO, International Institute for Educational Planning.

Harbison, F., & Myers, C.A. (1964). *Education, manpower, and economic growth: Strategies of human resource development.* New York: McGraw-Hill.

Inkeles, A., & Sirowy, L. (1983). Convergent and divergent trends in national education systems. *Social Forces* 62(2), 303-333.

Lazarsfeld, P.F. (1967). Some remarks on the typological procedures in social science. *Studies in Philosophy* 6(1), 119-139.

Lofland, J., & Lofland, L.H. (1984). *Analyzing social settings.* Belmont, CA: Wadsworth Publications.

McGinn, N. (1978). Types of research useful for educational planning. *Development Discussion Paper No. 24.* Cambridge, MA: Harvard University, Graduate School of Education, Center for Studies in Education.

McGinn, N., & Snodgrass, D. (1979). A typology of planning education for economic development. *Development Discussion Paper No. 62.* Cambridge, MA: Harvard Institute for International Development.

McKinley, J.C. (1966). *Constructive typology and social theory.* New York: Appleton.

O'Hare, M. (1989). A typology of governmental action. *Journal of Policy Analysis and Management 8*(4), 670-672.

Paulson, R.G. (1982). Choosing human resource interventions: A typology linking country conditions and educational development strategies. Draft report for discussion. *(Report No. SO 014706).* Pittsburgh, PA: Pittsburgh University, International and Development Education Program. (ERIC Document Reproduction Service No. ED 233 921).

Simon, H.A. (1983). *Models of bounded rationality: Vol. 2.* Cambridge, MA: MIT Press.

Spaulding, S. (1989). Comparing educational phenomena: Promises, prospects, and problems. In A.C. Purves (Ed.), *International comparisons and educational reform* (pp. 1-16). Virginia, Alexandria: Association for Supervision and Curriculum.

The Concise Columbia Encyclopaedia (1991). New York: Columbia University Press.

Thomas, L.G. (1968). Types of schooling for developing nations. Occasional paper of the international and development education clearinghouse. *(Report No. SO 004 891).* Pittsburgh, PA: Pittsburgh University, School of Education. (ERIC Document Reproduction Service No. ED 070 700).

UNESCO (1992). *World education indicators.* Paris: United Nations Educational, Cultural, and Economic Organization.

Walberg, H.J. (1989). Science, mathematics, and national welfare: Retrospective and prospective achievements. In A.C. Purves (Ed.), *International comparisons and educational reform* (pp. 99-111). Alexandria, Virginia: Association for Supervision and Curriculum Development.

Weiler, H.N. (1979). Notes on the comparative study of educational innovation. *(Report No. E 012 922).* Stanford, CA: Stanford University, California Institute for Research on Educational Finance and Governance. (ERIC Document Reproduction Service No. ED 192 444).

Appendix 4.1

Definitions of the World Education Indicators

Total population: Estimates of 1990 population, in thousands.

Population growth rate: Average annual percentage growth rate of total population between 1980 and 1990.

Population density: Number of inhabitants per square kilometre. Density figures are not given for areas of less than 1000 square kilometres.

Dependency ratios: Population in the age groups 0-14 and 65 years and over, expressed as percentages of the population in the age group 15-64.

Population in the 15-64 age group: Potential economically active population, i.e., population in the age group 15-64, in thousands.

Urban population: Number of persons living in urban areas, expressed as a percentage as a percentage of the total population.

GNP per capita: Gross national product per capita, in US dollars, and the average annual growth rate of GNP per capita between 1980 and 1989 in constant prices.

Illiteracy rate 1990: Number of adult illiterates (15 years and over) and those in the age group 15-19, expressed as percentages of the population in the corresponding age groups. Illiteracy is defined as all persons who have received less than four years of primary education or, in some cases, less than the historically and legally-specified compulsory-school duration.

Number of illiterates: Number of adult illiterates in 1990, in thousands, and the percentage change in the number of adult illiterates between 1970 and 1990.

Daily newspapers: Estimated circulation of general interest newspapers, expressed in number of copies per 1000 inhabitants.

Printing and writing paper: Consumption of printing and writing paper, other than newsprint, expressed in kilograms per 1000 inhabitants.

Radio and television receivers: Number of radios and television receivers per 1000 inhabitants.

Gross enrollment ratios, 4-23 year-olds: Total enrollment in pre-primary, first, second and third-level education expressed as a percentage of the population in the group aged 4-23 years.

Age group in pre-primary education: Population age group that according to the national regulations can be rolled at this level of education.

Gross enrollment ratios, pre-primary: Total enrollment in the education preceding the first level, regardless of age, expressed as a percentage of the population age group corresponding to the national regulations for this level of education.

Apparent intake rate, first level: Number of new entrants into first grade, regardless of age, expressed as a percentage of the population of official admission age to the first level of education.

Duration of compulsory education: Number of years of compulsory education, according to the regulations in force in each country.

Duration of first level education: Number of grades in primary education, according to the education system in force in each country in 1988.

Gross enrollment ratios / Net enrollment ratios: The gross enrollment ratio is the total enrollment in first level education, regardless of age, divided by the population of the age-group which officially corresponds to primary schooling. The net enrollment ratio only includes enrollment for the age group corresponding to the official school age of first level education.

Percentage of grade repeaters: Total number of pupils still enrolled in the same grade as the previous year, expressed as a percentage of the total enrollment at the first level.

Percentage of a cohort reaching grade 2, grade 4 and final grade: Percentage of children starting primary school who eventually attain the grade specified.

Coefficient of efficiency: This coefficient is the ratio between the theoretical number of pupil-years that it would have taken the graduates to complete the cycle of education, had there been no repetition or drop-out, and number of pupil-years actually spent by the cohort. The coefficient varies between 0 denoting complete inefficiency and 1 signaling maximum efficiency.

Transition from first to second level education: Number of new entrants into secondary general education, expressed as a percentage of the total number of pupils in the last grade of primary education in the previous year.

Duration of second level general education: Number of grades in secondary general education, according to the education system in force in each country in 1988.

Gross enrollment ratios / Net enrollment ratios: The gross enrollment ratio is the total enrollment in second level education, regardless of age, divided by the population of the age group which officially corresponds to secondary schooling. The net enrollment ratio only includes enrollment for the age-group corresponding to the official school age of second level education.

Pupil:teacher ratio: This ratio gives the average number of pupils per teacher at the level of education specified.

Percentage of female teachers: Number of female teachers, at the level specified, expressed as a percentage of the total number of teachers at the same level.

Teachers as percentage of non-agricultural economically active population: Total number of teachers in pre-primary, primary and secondary education, expressed as a percentage of the economically active population engaged in non-agricultural activities. The economically active population refers here to the population 10 years of age and over and covers all employed and unemployed persons.

Third level students per 100,000 inhabitants: Number of students enrolled at third level of education (or higher education) per 100,000 inhabitants.

Gross enrollment ratios: Total enrollment in education at the third level, regardless of age, expressed as a percentage of the population in the five-year age group following on from the secondary school leaving age.

University enrollment as percentage of total enrollment at the third level: Number of students enrolled in universities and equivalent degree granting institutions, expressed as a percentage of the enrollment in all institutions at the third level.

Teaching staff: Number of teachers in all institutions at the third level.

Percentage of students by field of study: Enrollment at the third level, in the broad field of study specified, expressed as a percentage of the total enrollment at the third level.

Percentage of female students in each field of study: Number of female students in each broad field of study, expressed as a percentage of the total enrollment (male plus female) in the field specified.

Field of study: Each field of study includes the following field, as defined by the International Standard Classification of Education (ISCED): education (science and teacher training); humanities (fine and applied arts; religion and theology; humanities); law and social sciences (law; social and behavioral sciences; commercial and business administration; home economics; mass communication and documentation; services trades); natural sciences, engineering and agriculture (natural sciences: engineering; mathematics and computer sciences, architecture and town-planning; transport and communications; trade, craft and industrial programs; agriculture, forestry and fisheries); medical sciences (medical sciences, health and hygiene).

Public expenditure on education as a percentage of GNP: Total public expenditure on education expressed as a percentage of the gross national product.

Public expenditure in education as percentage of government expenditure: Total public expenditure on education expressed as a percentage of total government expenditure.

Average annual growth rate in public expenditure on education: Average annual growth rates between 1980 and 1988 are based on the estimated total public expenditure in constant prices (data are deflated using the implicit GDP deflator) and have been computed by fitting trend lines to the logarithmic values of the expenditure data for each year of the period.

Current expenditure as a percentage of total: Public current expenditure on education, expressed as a percentage of total public expenditure on education.

Private enrollment as a percentage of total enrollment: Enrollment in private education, at the level specified, expressed as a percentage of the total enrollment at the same level.

Teachers' emolument as a percentage of total current expenditure: Expenditure on emoluments of teaching staff expressed as a percentage of total public current expenditure on education.

Percentage distribution of current expenditure by level: Public current expenditure by level, expressed as a percentage of total public current expenditure on education.

Current expenditure per pupil as multiple of GNP per capita: Public current expenditure per pupil, at each level of education, expressed in units of GNP per capita.

Chapter 5

Approaches to Setting and Selecting Achievement Standards

R. MURRAY THOMAS

University of California, Santa Barbara, USA

Each method of creating or choosing achievement standards is erected on a particular set of convictions about the functions that standards should serve. Those convictions are typically reflected in, first, a central principle that guides the development of the particular set and, second, a rationale explaining why that set is appropriate. The first purpose of this chapter is to review nine approaches to setting standards in terms of their guiding principles, supporting rationales, and potential personal-social consequences. The nine are identified by the following labels: (1) ideally-educated-person, (2) job-requirements, (3) available-openings, (4) normal-distribution, (5) compensatory-quota, (6) attainable-level, (7) individual-progress, (8) rationalized-combination, and (9) imprecise-intuitive. The chapter's second purpose is to describe four methods of selecting which source of standards to adopt in a given educational context. The four bear the titles: (1) imported-model, (2) accreditation-organization, (3) traditional-practice, and (4) political-pressure.

The phrase *achievement standards* has been used by educators with different meanings. Thus, it is important to explain that throughout this chapter the term, educational achievement standard, refers both to the *kind* of knowledge or skill students should learn and to *how well* they are expected to learn it. Furthermore, applying standards involves not only defining content and levels, but also specifying how learners' achievement will be assessed. Consequently, whenever a school is faulted for holding improper achievement standards, the criticism is usually aimed at:

1. What students are taught (curriculum content);
2. How well they are expected to learn it (level of mastery);
3. The methods used for assessing achievement (evaluation procedures).

The issue of curriculum content can be illustrated with a controversy in the United Kingdom regarding the teaching of English. In England, the National Curriculum Council prescribed in detail the aspects of grammar that children should display in their speech at each succeeding step of the schooling ladder. The plan was based on the conviction that "technically correct" speech should be the ideal on all occasions and that applying such a standard would benefit children of all social classes. In response to the National Council's prescription, the Curriculum Council for Wales produced a competing proposal that defined English standards in more general terms, emphasizing the influence of the varied contexts in which children speak and omitting such specifics as requiring correct use of pronouns and plurals in the primary grades. Both the National Council and Welsh Council listed authors whose works should be used as models of proper usage. The only author common to both lists was Shakespeare (Hofkins, 1993). In effect, the curriculum content that would comprise standards under the National Council differs substantially from that advocated by the Welsh Council.

The issue of mastery level is at stake whenever critics complain about the lack of a common standard behind the familiar symbols of achievement used in a nation's schools. For instance, the five-letter reporting system employed by most schools in the United States to signify performance levels have traditionally been defined with the following meanings:

A = superior achievement;
B = above average achievement;
C = average achievement;
D = below average achievement; and
F = failure.

However, critics contend that the level of student success behind a particular letter-grade can vary markedly from one school to another. Magaziner and Clinton (1992) contend:

Employers are aware that grading standards differ substantially across schools and courses. Consequently, most are skeptical of the usefulness and fairness of basing hiring decisions on grades received in school. This is one of the reasons that only one-fifth of the employers (in a survey of medium-sized business firms) requested a transcript or self-report information on grades in high school. One of the saddest consequences of this lack of objective information on young people's academic competencies is that employers with good jobs offering training and job security are unwilling to take the risk of hiring recent high school graduates. They prefer to hire workers with many years of work experience because their work records serve as a signal of competence and reliability (pp. 11-12).

The question of evaluation methods employed for judging achievement is reflected in a continuing debate about how well traditional examinations reflect students' mastery of learning objectives. Critics of examinations have contended that paper-pencil tests—as the dominant and oftentimes exclusive method of judging student progress—fail to measure many important educational outcomes and are not trustworthy predictors of vocational or social success (Broadfoot, 1992). To supplement or replace traditional examinations, advocates of broad-scale forms of evaluation recommend the use of such additional techniques as ratings of student work projects and performances, activity logs, social-participation records, student self-judgments, and the like. A variety of labels have been invented to identify such multifaceted modes of assessment—portfolios, activity record sheets, statements of achievement, letters of credit, biographical inventories, documented accomplishments, and profiles of achievement (Council of Europe, 1986; Eckstein & Noah, 1993; Knapp, 1992).

Most criticism of standards is aimed at curriculum content, the level of mastery, and assessment methods. However, sometimes the complaint is about the approach adopted for setting standards—in other words, about the conceptual foundations on which standards are constructed. This matter of approach is the concern of the present chapter. The chapter's first purpose is to explore conceptual foundations by defining nine approaches to setting standards, by identifying their underlying rationales, and by illustrating personal-social consequences that can result from each approach. A second purpose is to inspect four methods of selecting which type of conceptual foundation will be employed in a given educational system, identifying their supporting rationales, and suggesting personal as well as social outcomes that may accompany them.

Approaches to Setting Standards

For convenience of discussion, the nine conceptual foundations have been assigned these titles:

1. The ideally-educated-person approach;
2. The job-requirements approach;
3. The available-openings approach;
4. The normal-distribution approach;
5. The compensatory-quota approach;
6. The attainable-level approach;
7. The individual-progress approach;
8. The rationalized-combination approach; and
9. The imprecise-intuitive approach.

These types are not mutually exclusive. In practical educational settings, two or more are frequently combined.

As a preface to inspecting the nine options, it is useful to recognize that achievement standards are held at all levels of the educational enterprise. Parents hold standards for their children, and teachers set standards for their students. A common set of standards may also be held for an entire school or for all schools in a city, in a county, in a province, in a region, or in a nation. There can also be international standards, such as the set of expectations represented by the international baccalaureate, which is a certificate modeled after the French *baccalauréat* that is awarded to secondary-school graduates who pass a demanding examination covering specified curriculum areas (Eckstein & Noah, 1993, pp. 7-8). The approach to setting standards may differ from one of these demographic levels to another, so that problems can arise over which standard to apply to a given group of learners.

The Ideally-Educated-Person Approach

The Guiding Principle. It is possible to conceive of the type and level of knowledge and skill that would be displayed by a person who is ideally fitted to, first, participate constructively in society and, secondly, attain an optimal level of self-fulfillment. This type-and-level represents the standard against which all learners' achievement can be judged.

A Supporting Rationale. The central assumption underlying this approach is that the group or individual entrusted with setting standards is an authority in the particular field of study and thereby qualified to determine both the pattern of curriculum content and the levels of mastery (superior, good, passing, unacceptable, or the like) that the standard represents. Standards proposed by such an authority can be intended either for nationwide application or for only local use. Furthermore, the standards can be either obligatory (enforced in all schools) or optional (adopted by only those schools or instructors that choose to apply them). Examples of obligatory nationwide standards are those stipulated for all subject matter fields by the Ministry of National Education in France and in the People's Republic of China. Examples of optional national criteria are the standards recommended for science education in Australia and in the United States. The Australian Education Council (AEC), in 1993, offered a "final version" of "a framework for curriculum development in science education, setting broad goals and defining the scope and sequence of learning to be undertaken by every student during their schooling" (Australian Education Council and Curriculum Corporation, 1993, p. 2). In the United States, the American Association for the Advancement of Science (AAAS) in 1989 and again in 1993 suggested what students should know in science, mathematics, and technology by the time they complete grades 2, 5, 6, and 12 (Ahlgren, 1993; American Association for the Advancement of Science, 1989; Science class, 1993). Individual schools were free either to adopt or to reject the AAAS recommendations.

Potential Consequences. The ideally-educated-person approach works to people's general satisfaction whenever they agree that the proposer is truly qualified to set achievement criteria. However, the approach can be unacceptable to people who believe that the suggested curriculum content is unsuitable or that the levels of attainment are too high or too low. A familiar objection to curriculum content is the complaint that requiring students to pass tests in Latin and Greek is a waste of time in a modern world which finds little practical use for competence in "dead" languages. The objection that ideal levels of attainment are too high can be voiced when only small a proportion of learners earn acceptable marks, as often occurs in schools that enroll a high percentage of students from disadvantaged social backgrounds. In contrast, observers can complain that the ideal levels are too low whenever an excessively large proportion of students receive top-level scores. In recent years, such a complaint has become widespread in the United States.

> Harvard Magazine cites an unidentified dean of admissions at a top-[level] law school saying his office ignores *magna cum laude* and *cum laude* honors from Harvard because so may applicants have them. In 1993, 83.6 percent of Harvard seniors graduated with honors (Leo, 1993, p. 22).

Authors of the Australian science curriculum have been criticized for ostensibly manipulating the contents of their proposal so as to further an egalitarian social ideology that promotes the political agenda of feminists, environmentalists, multiculturalists, and socially disadvantaged minority groups. In short, a perennial debate continues over who is qualified to define the characteristics of the ideally educated person.

The Job-Requirements Approach

The Guiding Principle. An analysis of the tasks to be performed in an occupation or in a subsequent educational program for which the student is preparing should be the basis for defining the curriculum content and mastery levels that represent an educational program's achievement standards.

A Supporting Rationale. The most efficient use of a school's time and facilities will result if the curriculum focuses on the specific competencies students will need to succeed in a particular future occupation or educational program. Thus, the curriculum content should derive from the analysis of the knowledge and skills required for successfully carrying out the tasks of the target occupation or educational specialty. The minimum standard of attainment should be that level of mastery needed to succeed in the specialty at a beginning level.

Potential Consequences. The desirability of such an approach continues to be a matter of much controversy, especially the issue of curriculum content. Advocates of task analysis—particularly those who organize vocational-training programs—argue that providing learners with no more than a general education (communication skills, natural and social sciences) is a waste of time since it fails to equip students with the specific competencies they require to succeed in the years ahead. Proponents of national skill standards that must be met by all workers entering an occupation contend that such standards will improve a nation's "workforce and product quality in the global market place, provide better education and more portable employment credentials for workers, and increase accountability among schools, teachers, and vocational programs" (Hudelson, 1993, p. 32).

However, opponents of task analysis assert that in both primary and secondary schooling, a sound general-education curriculum is a more cost-effective investment than are vocational studies (Psacharopoulos & Woodhall, 1985, p. 314; White, 1988, p. 7). Difficulties associated with task-based programs include, first, rapid changes in the job market as a result of technological innovations and the shift of manufacturing jobs from one nation to another, second, the inability to predict accurately which students will eventually enter which occupations, and third, the high cost of providing the equipment and instructional expertise needed to prepare learners for specific vocations.

Actually, the fundamental issue is not whether specific vocational-skill standards are desirable; indeed they are, if a society is to operate efficiently. The issue, instead, concerns the question of which educational programs are best suited to perform the vocational-instruction role and hence to be governed by job-requirement standards. In other words, is the teaching of specific work skills most appropriately located in primary schools, in general secondary schools, in vocational secondary schools, in post-secondary centers, in apprenticeship programs, or where?

The Available-Openings Approach

The Guiding Principle. The available openings in an occupation or educational institution that learners wish to enter are limited to a given number. This number establishes the standard that separates the candidates who gain admission from those who do not.

A Supporting Rationale. An occupational field or an educational institution will operate most efficiently if it does not accept every applicant but, instead, accepts only the ones most likely to perform at a high level of effectiveness. This aim is best accomplished if there are more potential applicants than can be accommodated, if all applicants are evaluated for their aptitude to succeed in the occupation or educational program, and if the available openings are filled with the candidates who have received the highest evaluations. A French application of this widely adopted approach is the method used for enrolling students in the *École Normale* that prepared primary-school teachers. Candidates took a competitive examination that arranged them in an order of merit, and the predetermined number of places was distributed among those standing highest on the lists.

Potential Consequences. Proponents of this method contend that it makes economical use of training facilities by concentrating on those applicants who are most likely to succeed in the tasks ahead. Critics, however, note that predictions about which applicants will succeed are too often founded on untrustworthy information. Frequently, as in the *École Normale*, the selection derives from a single test. In other cases, a candidate's prospects are inferred from marks earned at preceding stages of schooling. However, a study in the United States revealed that six years after graduation there was only a 0.03 correlation between college grades (including test scores) and adult accomplishment. In another study, predictions of high-school graduates' potential as derived from a biographical inventory of their past school progress resulted in a 0.65 correlation between the inventory ratings and adult achievement (Knapp, 1992, p. 89). In short, predictions of who will perform well in a program that has a limited number of openings can be quite inaccurate.

The Normal-Distribution Approach

The Guiding Principle. Human traits are distributed in the pattern of a Gaussian curve, so it is proper to base the performance distinctions among learners on this curve.

A Supporting Rationale. In a randomly selected group of people, measures of such physical traits as height and weight show that the distribution of these characteristics across the group assumes the general shape of a Gaussian curve (named for the 19th-century German mathematician-astronomer Karl Friedrich Gauss). The largest number of people measure at the average; the farther that a measured height deviates from the average—either above or below—the fewer the people who will have achieved that height. Chance events, as in the tossing of dice or of coins successive times, also produce results that appear to match such a curve. Furthermore, the distribution of scores for such psychophysical skills as eye-hand reaction-time approximate the form of a Gaussian curve. In addition, students' scores on traditional tests of general intelligence, such as the Stanford-Binet Scale, assume a similar shape (Terman & Merrill, 1960, p. 18). Thus, because so many phenomena normally seem to be arranged in such a pattern, proponents of a normal-distribution-curve approach to setting standards have assumed that students' academic achievement levels should likewise normally be distributed in the same fashion. As a result, they deem it reasonable for educators to create measures of achievement that will spread students' scores in this normal pattern and to divide the resulting curve into consistent levels for the sake of reporting learners' success. Thus, adopting the conviction that achievement is distributed in the pattern of a Gaussian curve provides a statistically objective standard for reporting levels of performance.

Potential Consequences. Oftentimes individual instructors as well as school systems have applied the notion of normally distributed achievement as the basis for performance standards. For example, one university of my acquaintance stipulated that in any class containing a considerable number of students, the top 15 percent deserved a grade of A, the next 35 percent a grade of B, the next lower 35 percent a grade of C, and the bottom 15 percent either D or F, depending on how far they deviated from the group's average. Hills (1971, p. 698) cites an institution with a policy of dispensing letter grades in the following pattern—A=5 percent, B=20 percent, C=50 percent, D=20 percent, and F=5 percent.

The concept of a normal curve has also been applied to larger collections of students than those in a single class. The policy in Sweden for grading nationwide achievement tests is a case in point:

Marks are given from grade 8 on a five-point scale, the average award being 3. There are no specified percentages for the number of students receiving the differing grades, but the number of 1s and 2s should not normally exceed the number of 4s and 5s in a class. A 3 indicates the average achievement of all students in the country, but the average award for an individual class may be higher or lower than the national average. For the whole country the normal distribution of marks is shown below.

Mark	1	2	3	4	5
%	7	24	38	24	7

The mark received by any individual student should express to what extent he or she has succeeded in relation to the total population of students in the country taking the same subject. By means of a nationwide application of standardized achievement tests, it has proved possible to stabilize the marking system (Fägerlind, 1992, p. 83).

Critics of the normal-distribution approach have condemned it for both the underlying rationale and its likely personal-social consequences. In regard to the rationale, which contends that human events—including academic achievement—are normally distributed within any population, Hays has noted that:

There is nothing magical about the normal distribution; it happens to be only one of a number of theoretical distributions that have been studied and found useful as an idealized mathematical concept. Normal distributions do not really exist, however, and in applied situations the closest we can come to finding a normal distribution will never quite correspond to the requirements of the mathematical rule. Many concepts in mathematics and science that are never quite true give good practical results nevertheless, and so does the normal distribution (Hays, 1973, p. 296).

It should be recognized that the shape of a curve produced by test results is not entirely a function of students' levels of knowledge. The shape is also affected by the difficulty of the items composing the test. A markedly skewed distribution can result from a preponderance of either very difficult items or very easy items. Nevertheless, it is usually possible to arrange the difficulty levels of items so as to yield a result simulating a normal curve. However, this does not mean that levels of knowledge or skill within the group are necessarily distributed in such a manner. It should also be recognized that the task of determining where the cutting points should come in assigning letter grades or number grades is a subjective judgment and not a statistically objective phenomenon.

The question, then, is whether a normal-distribution approach does indeed, as Hays suggests, yield "good practical results" in setting achievement standards. The answer depends on what a normal-curve can and cannot accomplish. What such a standard does produce is a comparison of one student with another—sequentially ordering the members of the group from the most successful to the least successful on a given measure. However, the normal-curve approach does not reveal what proportion of a particular body of knowledge or skill any given student has mastered. Within a group in which all students have mastered nearly all of the learning objectives, one member who answered 85 percent of the test items correctly may turn out to have the lowest score and thereby receive the lowest mark—an American *F*, or a Swedish *5*. On the other hand, in a group in which none of the members have reached more than 15 percent of the learning goals, the student who has done least poorly (correctly answering 15 percent of the items) will receive a mark implying excellence—an American *A* or a Swedish *1*.

As Fägerlind noted in the Swedish case, when a large population of students serves as the foundation for a distribution curve, the mark a particular student receives is a more accurate indicator of that person's standing in comparison to others than when the curve is based solely on a small number of people, such as the students in a single class or in a single school.

The Compensatory-Quota Approach

The Guiding Principle. In any society, certain social and personal differences among people influence their access to schooling. Consequently, some people enjoy educational advantages that others are denied. In order to correct this imbalance and ensure that everyone has an equal chance for success, special opportunities need to be provided for disadvantaged individuals so that they can enter educational programs and graduate with recognized credentials. This means that standards for both enrolling and graduating less-advantaged candidates need to be different from the standards held for more-advantaged ones.

A Supporting Rationale. Several notable social and personal disadvantages influence whether people will enter educational programs and complete those programs in a respectable fashion. For example, members of upper socioeconomic classes are usually able to enter, and to succeed in, educational institutions more readily than are members of lower social classes. Parents in the upper social strata, as compared to those in lower strata, more often value advanced education, recognize what kinds of support are needed to promote their children's learning, provide such support, and know the political strategies needed to enhance their offsprings' chances of entering and graduating from particular programs. This tendency becomes increasingly pronounced the farther one progresses up the schooling hierarchy..

Other social factors also frequently affect access to schooling, such factors as people's ethnic, gender, religious, regional, or extended-family status. Within each of these types, certain groups have traditionally been accorded favored opportunities, causing the remaining groups to feel deprived of a fair chance to pursue their educational goals.

Not only can social factors place individuals at an educational disadvantage, but physical and psychological handicaps may have a similar effect. People who suffer sight or hearing disabilities, manual or locomotion disorders, mental retardation, or personality aberrations can be denied access to educational programs or, if admitted, can be left without adequate means to compete with schoolmates who are not encumbered by such burdens.

Consequently, if schooling is to be fair to all, then standards for accepting applicants and for graduating them from educational programs should be adjusted to the needs of the socially and physically and or psychologically disadvantaged. If the concept of fairness—of establishing "a level playing field"—is to be implemented, disadvantaged learners should not be held to the same standards as advantaged learners.

At the point of accepting candidates into a program, there are two principal ways to implement a compensatory policy. The first consists of two steps. First, inspecting the level of success that members of the disadvantaged group have achieved on entrance-requirement measures in the past (test scores, achievement profiles, prior school grades, letters of recommendation). Second, setting an entrance-requirement standard that will ensure the enrollment of a fair number from that group. It should be noted, however, that the term "fair" may be defined in various ways, including:

1. The percentage of entrants equal to the proportion of that social group in the general population;
2. Enough entrants to convince political activists that their group is being furnished adequate opportunities;
3. The number from that group who should have the types of skills the society needs.

The second way to provide special access involves establishing a quota system in which each disadvantaged group is assigned a given percentage of the available openings into the program. This percentage may match the proportion of the particular disadvantaged group that is found in the general population. However, as another option, the percentage may be some compromise figure that is less than the group's proportion in the population. Such a compromise may be forced on an educational institution by political pressure from competing groups who wish to enhance their own constituents' access to schooling. Or a compromise can also result from administrators' belief that the disadvantaged candidates will generally be of such poor quality that, if they are accepted in large numbers, they will lower the academic quality of the student body and damage the school's reputation.

Not only may entrance standards be reduced to accommodate the disadvantaged, but completion standards may be lowered as well. This can occur when candidates, who were originally accepted into a program by virtue of a reduced standard, subsequently fail to perform up to the level traditionally required for graduation. In this event, graduation standards may be reduced for such individuals so as to equip them with the symbols of achievement—certificates, diplomas, degrees, licenses—needed for entering desired occupations or further educational programs.

Potential Consequences. From the perspective of disadvantaged groups, one positive consequence of compensatory policies is that intellectually adept but socially disadvantaged individuals whose past educational preparation was of low quality will now have the chance to achieve to their potential. A second outcome is that compensatory provisions may help mollify political leaders who have complained that their constituents were not accorded ample educational opportunities.

However, a policy of applying lower standards to particular groups can also precipitate objections from people who do not qualify as "disadvantaged". Such objections surface when applicants with high entrance scores are denied admission and, at the same time, "disadvantaged" applicants with lower scores are admitted. The objectors argue their case by countering the compensatory principle with a meritocracy principle: "Applicants should be judged on their individual performance, not on their membership in a favored group".

Compensatory-quota practices may also result in broader societal consequences. If the reduced program-completion standards for the disadvantaged result in a significant number of such graduates later performing unsatisfactorily in their occupations, employers lose confidence in certificates and degrees as trustworthy indicators of competence. Furthermore, if graduates prove ill equipped to bear the responsibilities to which their credentials ostensibly apply, then the efficiency of the society suffers. Hence, a poor match between credentials and performance can distort or destroy the intended meaning of the credentials.

The Attainable-Level Approach

The Guiding Principle. The level of achievement that reasonably can be reached by the great majority of the learners, if not all of them, serves as the level of performance that participants are expected to attain.

A Supporting Rationale. The purpose of any educational program should be to enable all—or at least nearly all—of the participants in the program to master all of the learning objectives. Having a high proportion of the learners fail to achieve the objectives signifies that either instruction has been very poor, that the methods of evaluating student progress have been faulty, or that the achievement standards have been unreasonably strict.

Potential Consequences. The attainable-level approach has the advantage of enabling all learners to move ahead academically and feel that they have been a success. However, within a typical classroom or nation, students can differ significantly from each other in talent, in strength of motivation, and in the environmental conditions that foster academic progress. Such differences will be reflected in how well and how quickly members of the group master the learning goals. If all, or nearly all, students are to reach the same standard, then that standard must be set at a level which the least adept learners can reach and which the great bulk of the group can surpass. Thus, if the marking system only distinguishes between those who pass and those who fail, there is no way to indicate which students have barely passed and which ones have achieved far above the standard. Furthermore, if the content and pace of instruction have been set at a level easily attained by the least adept learners, then the average and above-average students will not be challenged to reach their potential, for they will lack the opportunity to learn far beyond the limits of such a program. In the public press, a critic of the "inflation of standards" in the United States complained that:

> For whatever reasons (and the feel-good self-esteem movement is surely one), marks have broken free of performance and become more and more unreal. They are designed to please, not to measure or to guide students about strengths and weaknesses. Give A's and B's for average effort and the whole system becomes a game of "Let's Pretend". Parents are pleased and don't keep the pressure on. Students tend to relax and expect high rewards for low output. What happens when they join the real world where A and B rewards are rarely given for C and D work? (Leo, 1993, p. 22).

One way that program planners have sought to deal with differences among students in learning potential has been to define more than one body of content (basic, extended, enriched) and several levels of attainment *(A, B, C, D)* that truly reflect different degrees of passing. To make such a system work, teaching methods must be adjusted to suit the individual differences in students' aptitudes.

The Individual-Progress Approach

The Guiding Principle. The standard for judging achievement is the amount of knowledge and skill an individual learner should acquire over a given period of time. This amount is computed as the difference between what the learner displayed at the beginning of the period and what he or she displayed at the end.

A Supporting Rationale. The participants in nearly any educational program will vary significantly in their inherited aptitude, their socioeconomic condition, their learning opportunities at home, their out-of-school responsibilities, and more. As a result, they enter the learning program with different levels of knowledge and skill. The purpose of the program should be to enable each student to progress from his or her present level to a reasonably higher level within a given period of time. In other words, the standard of achievement for every student should be set in light of that individual's level of attainment at the time of entering as well as that person's estimated potential for progress (aptitude, home conditions). In other words, the standard should be "fair" and "reasonable" in terms of the individual's potential.

Potential Consequences. The individual-progress method has been extolled as the mode of standard setting best suited to promoting the mass-education goal of "Let each become all that he or she is capable of being". However, the approach has been criticized on several counts. Observers have noted that comparing a student's progress only against his own past record can give the less apt learners a false sense of their actual ability. Consequently, after receiving high grades for their modest progress in school, they receive a rude shock in the world beyond school when they fail to compete successfully with their more capable age mates. Furthermore, parents and employers typically wish to know, not just how a youth measures against his or her own estimated potential, but also how that young person compares to others the same age. An individual-progress approach fails to yield this sort of information.

There is also a school-wide version of the individual-progress method. Achievement standards within a school may be adjusted to the apparent capabilities of the particular student population. In a socially depressed ghetto the distribution of marks within a government school's population may be quite similar to the distribution in a private school attended by a highly select population of socially privileged students. A letter grade of A or of C in the two schools can have quite a different meaning in terms of academic accomplishment.

The Rationalized-Combination Approach

The Guiding Principle. A logically rationalized combination of approaches furnishes a better balanced foundation for achievement expectations than does any single method of setting standards.

A Supporting Rationale. Each method of setting standards is accompanied by strengths and weaknesses. The best way to take advantage of the strengths and to diminish the weaknesses of the various options is to devise a well-reasoned combination of two or more app-roaches.

Such a rationale often results from the fact that achievement standards, other than those established within an individual classroom, are rarely decided by a single person. Instead, they are determined by a committee or a series of committees whose members can hold varied opinions about the best way to set standards. As a consequence, the final product of their deliberations can be an integration of two or more approaches, a compromise intended to satisfy various constituencies. The resulting standards are then publicly defended by a line of logic intended to convince observers of the virtues of the ultimate product.

Potential Consequences. An advantage of this method is that it may prevent attacks on the standards by politically significant dissidents. A disadvantage is that inconsistencies within the ultimate set of standards may render them difficult to apply. For example, socially disadvantaged ethnic or religious minorities within the society may insist on a quota system by which members of their group will be admitted to educational programs in the proportion that their group represents in the general population. The putative purpose is to compensate their group for a lack of educational opportunities in the past. At the same time, representatives of nondisadvantaged groups may contend that individual candidates should be judged solely on their demonstrated achievement, regardless of their background. Negotiating an acceptable compromise between these two positions can be a difficult task (Thomas, 1983).

The Imprecise-Intuitive Approach

The Guiding Principle. It is important to recognize the key role that intuition plays in all decision making, including the decision about what type of achievement standard is most suitable in a given situation.

A Supporting Rationale. No decision making is solely a function of conscious logic. Instead, decision making always includes a measure of intuition that reflects unconscious motives, emotions, and the subliminal calculation of unrecognized but influential factors. The attempt to ignore thought processes that operate below the threshold of consciousness is simply self-deception. Therefore, in addition to engaging in logical analysis when setting standards, we should be willing to include our intuitive "sense of what is right" and "feeling of what will work". That is the soundest way to make sure that both conscious and unconscious factors are being accommodated.

Potential Consequences. A problem with depending on intuition is that it rules out rational discussion and reasoned negotiation among the people setting the standards. Accepting standards founded on someone else's intuition results either from blind faith in the person who issues the standards or from fear that such a person may retaliate against those opposing the proposal.

Approaches to Selecting a Type of Standard

Not only are there different conceptual bases for setting standards, there are also different ways of choosing which type of standard to adopt. The four selection approaches described below are identified by the following labels:

1. The imported-model approach;
2. The accreditation-organization approach;
3. The traditional-practice approach;
4. The political-pressure approach.

The Imported-Model Approach

The Guiding Principle. The type and level of standards that an admired educational system uses will be an appropriate type and level for one's own system to adopt.

A Supporting Rationale. Typically, whenever an individual school or an entire nation imports its achievement standards from an esteemed school system, it is on the assumption that the success the foreign system enjoys is due—at least partly—to the standards it holds.

Potential Consequences. There are two principal circumstances that lead to the importation of standards. The first involves a colonial power bringing standards from their homeland to impose on schools in the territories they hold as colonies, because colonialists consider the indigenous people's standards inadequate in content and level of attainment (Thomas, 1981).

The second circumstance finds the educational authorities who control a school system voluntarily adopting standards from outside their school system. Two conditions necessary for such an occurrence are, first, dissatisfaction with the achievement of one's own students and, second, the identification of a source of standards that presumably would lead to more acceptable outcomes. These two conditions are often produced by the same event. For instance, in recent decades a series of assessments have been conducted of the academic achievement of students in several dozen nations. The most prominent of these cross-national appraisals have been sponsored by the International Association for the Evaluation of Educational Achievement (IEA) which develops measures of student progress suited to the curricula of the participating countries; the curriculum areas attracting the greatest interest have been those of mathematics, science, reading, and writing (Degenhart, 1990; Elley, 1992; Postlethwaite & Wiley, 1992; Keeves; 1992).

A chief advantage of importing standards is that school authorities can assert that their students meet the same criteria as those held within the original source of the standards. In effect, students succeeding in the system bear credentials equivalent to the credentials of students in the schools from which the standards were derived. The chief disadvantage is that such standards, in terms of content, may be poorly suited to the needs and conditions of the societies that are the recipients of the imports.

In speaking of colonized regions of Africa, Coombs (1985, p. 106) has maintained that as schooling developed in those areas, it became "increasingly maladjusted in relation to the real learning needs of students and the development needs of their changing societies".

The Accreditation-Organization Approach

The Guiding Principle. The achievement standards in our school should be those advocated by a reputable association or board that certifies the institutions as being of high quality.

A Supporting Rationale. In order to ensure that the education provided be well regarded by other institutions and the public, evidence must be provided to show that the standards are the same as those endorsed by a prestigious organization that attests to the high distinction of institutions deserving certification.

Accreditation organizations can operate under the auspices of various kinds of sponsors, including a ministry of education, a provincial education department, a religious order, a coalition of schools and universities within a region or nation, a professional association (medical, engineering, legal, architectural, and the like), or an international educational association. Such bodies have the following functions:

1. Establishing criteria-of-excellence to be used in assessing all aspects of an educational institution's operation;
2. Conducting on-site evaluations of how well a given institution meets the criteria;
3. Suggesting reforms to be instituted in any aspects of the operation that warrant improvement; and
4. Signifying approval of the institution by according it official accreditation.

Potential Consequences. One frequent result of an education system accepting accreditation standards is that the managers of the system are motivated to maintain the quality of facilities, staff, services, and achievement criteria demanded by the accreditation body. Furthermore, graduates of accredited schools enjoy the prestige associated with the accreditation body. However, a school's commitment to the requirements of a certifying organization may unduly restrict the flexibility the staff requires for creating new options to accommodate social change. In other words, accreditation bodies tend to conserve existing policies, ones that may prove poorly suited to modifications in the school's mission or clientele.

The Traditional-Practice Approach

The Guiding Principle. The standards that have been used in the past in our education system should continue to be used in the future.

A Supporting Rationale. Adopting this approach can result from any of several motives—satisfaction, resigned acceptance, or fear.

Satisfaction is the motive whenever people are pleased with the present outcomes of the education system. They defend their position with such admonitions as "It's not wise to change just for the sake of change" and "If it ain't broke, don't fix it". They contend that traditional standards have stood the test of time, have produced successful graduates, and have furthered the progress of the society.

Resigned acceptance reflects people's belief that the traditional standards are so deeply entrenched that it would be futile to try instituting something new. This apathy is usually seated in the conviction that the educational bureaucracy (policymakers, administrators, teachers, support personnel) would oppose significant change and that the citizenry is too accustomed to the present practices to advocate a departure from tradition.

Fear is the motive whenever people resist adopting a new standard because "You really can't be sure how it would work. It might make things a lot worse. Introducing such a plan wouldn't be worth all the trouble it could cause".

Potential Consequences. An advantage of continuing to use traditional standards is that people already understand what those standards are and how they work. Such is not the case with the adoption of a new departure that is unproved in the present setting. In short, there is often greater comfort in the known than in the yet untried. On the other hand, traditional standards—like imported ones—may fail to suit changes in instructional technology, in the labor market for which students are preparing, in the kinds of students attending the schools, or in social reforms the schools are expected to promote.

The Political-Pressure Approach

The Guiding Principle. An achievement standard should be chosen that will be acceptable to individuals or groups which, if dissatisfied, could impose detrimental sanctions on the educational establishment. In other words, the standard should not invite punitive reactions from important political constituencies.

A Supporting Rationale. If an education system is to earn the respect and support of the public it serves, then its standards of achievement must be well regarded by that public. When powerful constituents disagree with a type of standard, they can damage the educational establishment by discrediting or ousting its personnel, withholding financial support, and diverting able students into other educational programs.

Potential Consequences. Which political pressures will prove significant in a given case of standards depends upon a variety of factors, including the degree of satisfaction of different political factions, the administrative level at which standards are set, and the organizational sophistication and size of pressure groups.

Degree of Satisfaction. In multicultural societies, the likelihood that ethnic groups will lobby for changes either in educational services or in the application of standards is partially determined by how well members of the group are succeeding under present standards. A study conducted in Britain during the 1980s led the authors to conclude that:

> The generally received opinion is that the 'problem' of the educational attainment of children from ethnic minorities amounts to this: newly arrived immigrants tend to find it very difficult to adjust to British schools, but the longer they have lived in this country, the better they cope. 'Asian' children, indeed, soon come to perform at a level indistinguishable from the white majority. But children of West Indian origins continue to lag behind the indigenous majority. Children of Pakistani origin, it appears, are now obtaining rather lower scores on standard tests of attainment and of IQ than are West Indian children (Mackintosh, Mascie-Taylor & West, 1988, p. 98).

Under such conditions it is more likely that pressure to change services and standards will be exerted by the West Indian and Pakistani communities than by 'Asians'.

Administrative Level. When standards are set at a nationwide level, the most effective efforts to determine them are more often conducted by large, well financed organizations.

> In the United States ... the National Parent-Teacher Association has endorsed the need for national standards and means to assess and implement them, though it has gone on record against several practices: a school curriculum driven by testmakers, 'top-down' reforms dictated by national authorities, and multiple-choice examinations. ... The French *Fédération des parents d'èléves de l'enseignement public* [parents of public-school pupils] has grown in size and visibility, pressing for reform in the content and structure of the *baccalauréat* (Eckstein & Noah, 1993, p. 97).

However, when standards are set at the local school or community level, even individuals and small parent groups can often successfully exert political pressure.

Organizational Sophistication. The greater ability of some groups to influence the choice of standards derives from their superior understanding of how the political system operates and from their skill in devising strategies that convince decisionmakers to accede to their demands. Groups composed of wealthy, well-educated, indigenous citizens are typically more cohesively organized and better positioned to influence decisionmakers than are newcomers to the society who have neither the education nor funds to mount a significant campaign expressing their concerns about standards—if, indeed, they are even aware of the present standards and of alternatives to propose.

Conclusion

A lesson that might be inferred from the material in this chapter is that people who select achievement standards will be better equipped to rationalize their choices if they recognize the assumptions underlying alternative approaches and understand the strengths and weaknesses of those alternatives. As a practical application of this lesson, people who advocate a particular approach can profitably accompany the presentation of their proposal with a rationale that answers two questions:

1. *Who profits from the recommended approach and why?* Certain approaches to setting standards promote the welfare of some groups or individuals over the welfare of others. In short, not everyone profits equally from any given approach. Thus, advocates of a particular type of standard may find it judicious to include in their presentation a description of the entities that profit most from their proposal and an explanation of why the well-being of those entities warrants favored consideration.

2. *Who will likely object to the approach and why?* Critics of a given approach to standards typically cite the deleterious effect the plan will have on selected groups or individuals. Thus, in anticipation of such criticism, proponents of the approach may find it prudent to forestall attacks by including in their published proposal a discussion of likely objections to their plan and an explanation of why the advantages of their system outweigh its ostensible disadvantages.

References

Ahlgren, A. (1993). Creating benchmarks for science education. *Educational Leadership* 50(5), 46-49.

American Association for the Advancement of Science. (1989). *Science for all Americans.* Washington, DC: Author.

Australian Education Council and Curriculum Corporation. (1993). *A national statement on science for Australian schools.* Canberra: Author.

Broadfoot, P. (1992). Toward profiles of achievement: Developments in Europe. In M.A. Eckstein & H.J. Noah (Eds.), *Examinations: Comparative and international studies* (pp. 61-78). Oxford: Pergamon Press.

Coombs, P.H. (1985). *The world crisis in education: The view from the eighties.* Oxford: Oxford University Press.

Council of Europe (1986). *Assessment and certification: Issues arising from the pilot projects.* Strasbourg: Author.

Degenhart, R.E. (Ed.) (1990). *Thirty years of international research: An annotated bibliography of IEA publications.* The Hague: International Association for the Evalua-tion of Educational Achievement.

Eckstein, M.A., & Noah, H.J. (1993). *Secondary schools examinations: International perspectives on policies and practice.* New Haven, CN: Yale University Press.

Elley, W.B. (1992). *How in the world do students read?* The Hague: International Association for the Evaluation of Educational Achievement.

Fägerlind, I. (1992). Beyond examinations: The Swedish experience and lessons from other nations. In M.A. Eckstein & H.J. Noah (Eds.), *Examinations: Comparative and international studies* (pp. 79-87). Oxford: Pergamon Press.

Hays, W.L. (1973). *Statistics for the social sciences.* New York: Holt, Rinehart, & Winston.

Hills, J.R. (1971). Use of measurement in selection and placement. In R.L. Thorndike (Ed.), *Educational measurement* (pp. 680-732). Washington, DC: American Council on Education.

Hofkins, D. (1993). England rejects the prints of Wales. *Times Educational Supplement,* April 23, p. 14.

Hudelson, D. (1993). The standard approach. *Vocational Education Journal 68*(2), 24-32, 51.

Keeves, J.P. (Ed.) (1992). *The IEA study of science III: Changes in science education and science achievement: 1970 to 1984.* Oxford: Pergamon Press.

Knapp, J. (1992). Commentary. In M.A. Eckstein & H.J. Noah (Eds.), *Examinations: Comparative and international studies* (pp. 88-91). Oxford: Pergamon Press.

Leo, J. (1993). 'A' for effort. Or for showing up. *US News and World Report 115* (15), 22.

Mackintosh, N.J., Mascie-Taylor, G.G.N., & West, A.M. (1988). West Indian and Asian children's educational attainment. In G. Verma & P. Punfrey (Eds.), *Educational attainment* (pp. 87-99). London: Falmer Press.

Magaziner, I., & Clinton, H.R. (1992). Will America choose high skills or low wages? *Educational Leadership 49*(6), 11-14.

Noah, H.J., & Eckstein, M. A. (1992a). Comparing secondary school leaving examinations. In M.A. Eckstein & H.J. Noah (Eds.), *Examinations: Comparative and international studies* (pp. 3-17). Oxford: Pergamon Press.

Noah, H.J., & Eckstein, M.A. (1992b). The two faces of examinations: A comparative and international perspective. In M.A. Eckstein & H.J. Noah (Eds.), *Examinations: Comparative and international studies* (pp. 147-170). Oxford: Pergamon Press.

Postlethwaite, T.N., & Wiley, D.E. (1992). *The IEA study of science II: science achievement in twenty-three countries.* Oxford: Pergamon Press.

Psacharopoulos, G., & Woodhall, M. (1985). *Education for development: An analysis of investment choices.* New York: Oxford University Press.

Science class (1993, October 25). *San Luis Obispo Telegram-Tribune,* p A3.

Terman, L.M., & Merrill, M.A. (1960). *Stanford-Binet intelligence scale.* Boston: Houghton Mifflin.

Thomas, R.M. (1981). Evaluation consequences of unreasonable goals—The plight of education in American Samoa. *Educational Evaluation and Policy Analysis 3*(2), 41-50.

Thomas, R.M. (1983). Malaysia. In R.M. Thomas (Ed.), *Politics and education: Cases from 11 nations* (pp. 149-169). Oxford: Pergamon Press.

White, M. (1988). Educational policy and economic goals. *Oxford Review of Economic Policy 4*(3), 1-20.

Chapter 6

The Validity and Reliability of Outcome Measures

RICHARD M. WOLF

Teachers College, Columbia University, New York, USA

Validity and reliability are key concepts in educational measurement. The latest edition of the Standards for Educational and Psychological Testing *(AERA, APA, NCME, 1985) list validity and reliability as the first two standards to be considered in judging the adequacy of measures. This is no accident. Educational studies that use measures lacking in validity and reliability will produce worthless results regardless of how well sampling, data collection and analysis are carried out. Careful consideration of these two key concepts are necessary for anyone who uses educational and psychological measures whether it be the testing of an individual student or a large-scale evaluative study. This chapter addresses these two key criteria.*

While the focus of this chapter is on the validity and reliability of outcome measures, these concepts have broad applicability. The Paris-based Organization for Economic Cooperation and Development (OECD), in its recent publication *Education at a Glance: OECD Indicators* (1993), has classified educational variables into four broad categories. These are:

1. Context variables—demographic, social and economic factors, etc.;
2. Input variables—money, human resources, buildings, etc.;
3. Process variables—enrollments, decision-making, opportunity to learn, implemented curriculum, etc.; and
4. Outcome variables at the *student* level (mainly achievement in key subjects), outcomes at the *system* level (between and within school variation in achievement, graduation rates, science degrees, etc.), and

labor market outcomes (unemployment and labor force participation rates, educational attainment of the population, education level in industry, education and earnings variation).

Issues of validity and reliability pertain to all four of these categories although they are infrequently studied in the first three. When they have been studied, serious issues regarding validity and reliability of measures often arise (Wolf, 1993). Thus, while this chapter is limited to the validity and reliability of outcome measures, the concepts apply to all four categories of educational variables.

Validity

Validity is concerned with whether an instrument is measuring what it is supposed to measure. The concept refers to "... the appropriateness, meaningfulness, and usefulness of the specific inferences made from test scores" (Standards, 1985, p. 9). Obviously, if a test is not appropriate, meaningful or useful, then any inferences made from its use are, at best, questionable and more likely, irrelevant or downright wrong. There are three recognized forms of validity: content validity, criterion-related validity and construct validity. Of the three, construct validity is the most basic since it addresses the fundamental question of what is the meaning of a score on a measure.

Construct Validity

While construct validity pertains to all types of measures, it is especially critical in judging the adequacy of affective or personality measures where content validity is deemed inadequate and criterion-related validity is usually not possible because of the absence of any criterion measure as a standard against which to judge the test. Construct validity entails a number of studies, usually conducted over a period of time to establish the validity of a measure. The two principal kinds of studies that are conducted to establish the construct validity of an instrument are studies of group differences and studies of correlations with other tests. Studies of group differences involve administering a new instrument to groups who are expected, on the basis of theory or previous research, to score at different levels on an instrument and to compare the obtained results with prior expectations. For example, an attitude measure might be administered to students in regular, remedial, and accelerated sections of a particular course to determine if the favorableness of student attitudes towards a course differs by section. Similarly, an attitude scale measuring attitudes on a given social issue could be administered to members of groups known to have differing public positions on the issue. Studies of correlations with other tests are undertaken to determine if a new instrument has expected relationships with other previously established measures. If results accord with expectations, then evidence is accumulated towards demonstrating the construct validity of the instrument.

Criterion-related Validity

Criterion-related validity is the primary means of establishing the validity of ability and aptitude measures. In a criterion-related validity study, the prospective instrument is administered to all applicants to an institution or for a particular job, the results are locked away and, later, a measure of performance in the situation, referred to as the criterion, is obtained. Scores on instruments are then related to the criterion and the resulting correlation coefficient is termed the validity coefficient.

One of the biggest problems one faces in determining criterion-related validity of an instrument is the selection of a suitable criterion measure. What, for example, would be a suitable criterion of academic achievement for an academic aptitude test? The answer is not self-evident. There are four considerations in the selection of a criterion measure. First, the criterion must be relevant. This is a matter of judgment and different people might have different ideas as to what is a relevant criterion. Second, a criterion measure must be free from bias. That is, it must be free from external factors that might bias the measure. Third, a criterion measure must be reliable. It must be reasonably accurate and precise. Fourth, a criterion measure must be available. One must be able to obtain it. Finding a criterion measure that meets these four considerations is not an easy task. The usual criterion measure for an academic aptitude measure is the grade point average obtained in an educational program. While this is usually available, it is not without problems. The grade point average is often not a pure measure of achievement. It often contains some measure of industry on the part of the student along with other non-cognitive factors. Since grading standards often vary between teachers and between different subjects areas, it is often difficult to know exactly what the grade point average actually represents. However, it is one of the most frequently used criterion measures of academic aptitude because of its ready availability.

Content Validity

Content validity is the major form of validity used with measures of educational achievement. Content validity is judgmentally determined by reviewing test items and tasks in relation to the specifications set forth in a test plan. The test plan consists of, among other things, the content topics, skills and processes covered in instruction. Test developers use the test specifications to produce the items, problems, and tasks that constitute a test. Once this is done, specialists in a subject area are called upon to review the test material in terms of the test specifications. This process is called content validation. It is the critical step in the determination of content validity.

Content validity can never be assumed. It must be carefully determined. There are a number of reasons for this. Well-intentioned test developers and item writers may feel that they have constructed a test according to the test specifications but may not have done so. For example, if a test of science achievement consists of difficult prose, the resulting test may be more a measure of reading comprehension than of science achievement. If this happens, then scores on the resulting test may be less of an indication of the achievement in science of an examinee than of his or her reading proficiency. For this reason, test developers are encouraged to use vocabulary that is somewhat below the reading level of the student for whom the test is intended. Grammatical cues can also invalidate items on achievement tests. Consider the following test item:

An erb is most like a:

A. ute
B. oon
C. iro
D. bim

Examinees with no knowledge of what is being tested, but with some knowledge of grammar can obtain the correct answer to this nonsensical item by recognizing that, in the English language, nouns that start with a vowel are preceded by "a" while nouns that begin with a vowel are preceded by "an". Thus, many students could correctly select "bim" (D) as the correct answer although the item is pure nonsense. This flaw, incidentally, can be easily corrected by revising the stem of the item as follows: "An erb is most like a (an)".

Many other examples of deficiencies in the presentation of test questions can, unfortunately, be cited. Nothing is likely to be gained by reciting a list of flaws. Examples of ways of writing test questions so that they are free from flaws are provided in standard texts (Gronlund & Linn, 1990, pp. 177-189). The point is to construct test questions so that examinees can obtain the right answer if they possess the information and skills to answer a question, but will not obtain the correct answer if they do not. The principle is clear and straightforward, but achieving it in practice is often difficult. Careful review of test questions not only in terms of test specifications but also in terms of the criteria for good test items is needed to insure content validity of achievement measures.

While content validity is judgmental in nature and involves reviewing each test question or item in relation to test specifications, several indices have been developed to quantify aspects of content validity (Rosier & Keeves, 1991, pp. 299-305). The first index is the *test coverage index*. This index provides an estimate of the extent to which a particular curriculum has been covered by the questions or items included in a test. The higher the numerical value of this index (which has a maximum of 1.0), the greater the amount of coverage of a curriculum. The second index, the *test relevance index*, provides a measure of the degree to which the questions or items included in a test were taught through a particular curriculum or were relevant to that intended curriculum. The third index, the *curriculum coverage index*, measures the extent to which a school system covered the more general curriculum, defined by the body of content that might be taught or was taught in other school systems within a country or in other countries. Again, the higher the numerical value of the index, the greater the coverage. The curriculum coverage index serves as a guide to how much of a common body of knowledge is being taught in a particular locale, whether it be a single school or even an entire country.

Use of the three indices described above can help educators judge various aspects of the content validity of achievement measures. Unfortunately, there are no fixed values for any of the indices at which one cay say that a particular test's content validity can be called into question. Keeves and Rosier (1991) do, however, provide some useful general guides as well as a number of examples of the application of these indices.

Conclusion

Validity is the key criterion for judging the adequacy of outcome measures, whether they be in the area of achievement, attitudes, or aptitudes. Measures that are lacking in validity will yield useless or, worse, misleading information.

Reliability

Reliability is concerned with the accuracy or precision with which a characteristic is measured. Some characteristics such as height and weight can be measured with a high degree of accuracy. Other characteristics are measured with much less accuracy. Clearly, the more precisely a characteristic can be measured, the greater the reliance one can place on a score obtained with a measure. Reliability is sometimes confused with validity. This is unfortunate since the two are quite different. Validity is concerned with whether an instrument is measuring what it is supposed to measure while reliability is concerned with the precision of measurement. An instrument can have high reliability without having any validity. For example, a measure of height will be extremely reliable (measured with high precision) but may have no validity whatsoever as a measure of, say, science achievement. Reliability is necessary for validity, but not sufficient.

The concept of reliability was developed in connection with instruments to measure characteristics of individuals. That is, reliability has been traditionally concerned with the precision of test scores of individuals. This, of course, is critical when one wishes to make statements about the ability and achievement performance of individual students. Test reliability is somewhat less of a concern when one wishes to make statements about the performance of groups of students. This point will be elaborated later. It may also be noted that reliability also applies to other measures such as home background characteristics of students and teacher and school characteristics.

The reliability of a measure is expressed by a reliability coefficient. This is usually a correlation coefficient but is termed a reliability coefficient because it expresses the relationship of a test with itself. There are three ways in which the reliability of a measure can be estimated. These are: (1) alternate form, (2) test-retest, and (3) internal consistency.

Alternate Form Reliability

Alternate form reliability for a measure is established by administering an instrument to a group of examinees and then administering a parallel or alternate form of the measure to the same group of examinees. The second administration may take place on the same day or a few days later. The resulting correlation coefficient is referred to as the alternate or parallel form reliability of the measure. It is the most stringent procedure for establishing reliability, especially if there is an interval between the first and second testings, since it takes into account variations in the sample of items administered, day to day variations, and variations in the individual's speed of work. If one obtains a satisfactorily high reliability coefficient using this approach, one can be quite confident of a measure's reliability.

Test-retest Reliability

Test-retest reliability involves re-administration of the same test to the same group of examinees with or without an interval between testings. The correlation coefficient between the two sets of test scores is a test-retest reliability coefficient. Test-retest reliability is often used with tests in which speed of response is an issue. For example, a manual dexterity test involving moving pins, one at a time, from one dish to another with a pair of tweezers within a very short period of time, say two minutes, can be used to establish the reliability of the measure. The correlation coefficient between the number of pins moved on each occasion will give the test-retest reliability of the task.

Internal Consistency Reliability

Both alternate form and test-retest reliability require having the same group of examinees sit for two testings. Often this is impractical. For this reason, a variety of procedures have been developed to estimate the reliability of a test based on a single administration. These approaches all fall under the heading of internal consistency reliability.

Three of the most widely used internal consistency approaches to estimate reliability are : (1) split-half reliability, (2) Kuder-Richardson Formula 21, and (3) Kuder-Richardson Formula 20 and its generalized form, coefficient alpha. Each of these yields a reliability estimate based on a single administration of a test.

Reliability Thresholds

Since the reliability coefficient of a measure is a correlation coefficient, these can be compared across measures to determine which is the most reliable. Obviously, measures with higher reliability coefficients are preferred over measures with lower reliability coefficients, given that the validity of the measures has been established. However, the question that has not been addressed is how high should the reliability of a measure be in order for it to be considered acceptable for routine use? Well-made aptitude and achievement tests will often have reliability coefficients of 0.9 or higher. Some achievement measures with somewhat heterogeneous content will have reliability coefficients of 0.8 or higher. Personality measures often have reliability coefficients of 0.75 or higher. As a rule, one should not use any test for individual measurement with a reliability coefficient below 0.7. The reason for this will be presented below.

Once one has selected a measure for use, the reliability coefficient is less important than the standard error of measurement. The standard error or measurement describes the amount of precision that can be accorded an individual score. It represents an average amount of measurement error for individual scores and is obtained using Eqn 6.1:

$$se = sd \sqrt{1\text{-}rel} \qquad\qquad\qquad\qquad \text{Eqn 6.1}$$

where:
se = standard error of measurement
sd = standard deviation of group taking the test
rel = reliability coefficient of the test

Information about the standard error of measurement can be used to construct Table 6.1, which sets forth the relationship between test reliability and the standard error of measurement.

Table 6.1 shows that, as the reliability coefficient increases, the standard error of measurement decreases. If the reliability coefficient is 0.7, the standard error of measurement will be slightly larger (.55) than one half of a standard deviation for the group. Clearly, this is a large amount of error. Hence, the statement that educators should not use measures that have reliability coefficients below 0.7 for individual measurement. There is simply too much measurement error attached to each individual's score.

Table 6.1. Relationship between test reliability and the standard error of measurement

Reliability coefficient	Standard error of measurement
0.5	0.71 SD
0.6	0.63 SD
0.7	0.55 SD
0.8	0.45 SD
0.9	0.32 SD
0.95	0.22 SD
0.98	0.14 SD

The above discussion is only part of the story with regard to the reliability of measures. While an instrument with a reliability coefficient below 0.7 may be inadequate for describing the performance of an individual, the instrument may be serviceable for describing the performance of groups. The reason for this is that measurement errors tend to cancel one another out when one averages scores for a number of examinees. In fact, the basic assumption regarding measurement errors is that they are normally distributed with a mean of zero. Thus, averaging scores for a group of examinees will tend to cancel out the errors and yield an estimate of the performance of a group that will be reasonably accurate, if the group is reasonably large. The average of a class group of twenty to twenty-five students will thus be a good estimate for the performance of the group.

The point of this discussion is that one can use measures with somewhat lower reliabilities to describe the performance of groups than one would be able to use to describe the performance of individuals. In practical terms, measures with reliabilities as low as 0.5 can be used to describe the performance of class groups. If even larger units, such as entire countries, are to be described, the reliabilities can be even lower, e.g., 0.4, although one should avoid using them if possible.

One note of caution must be stated. While one can use measures with somewhat lower reliabilities to describe the performance of groups as opposed to individuals, one cannot then obtain correlations at the level of the individual between somewhat unreliable outcome measures and other characteristics of individuals such as home background and attitudes. Such an action would assume that the measures are reliable at the individual level.

Factors Influencing Reliability

There are several factors known to affect the reliability of measures. They are: (1) range of the group being tested; (2) level of proficiency of the group; (3) length of the measure; and (4) procedure used to estimate reliability. Each will be commented on in turn.

The reliability coefficient indicates how consistently a measure places an individual in relation to others in the group. When there is little shifting from one form to another, the reliability coefficient will be high. But the extent to which individuals will switch places depends on how similar they are. It does not take very accurate testing to distinguish the reading ability of nine year olds from that of thirteen year olds. But to place each nine year old accurately within a nine year old group is considerably more demanding. Thus, the more homogeneous the group, the lower the reliability coefficient will be. Conversely, the more heterogeneous a group, the higher the reliability coefficient will be if other things are equal.

The level of proficiency in the group being tested will also influence the reliability of a measure. However, no simple rule can be formulated for stating the nature of this relationship since it will depend to a large extent on the way in which the measure was built. For those individuals for whom the measure is very difficult and do a large amount of guessing, accuracy is likely to be low. Similarly, when the test is so easy that a group can correctly answer most of the items easily, it may be expected to do a poor job in distinguishing among the group members. Generally, a test will measure somewhat more accurately for those individuals who score near the center of the distribution than those who succeed with only a few items or with almost all of them. In effect, a longer test is operating for individuals in the middle range rather than those at the extremes and this results in a more accurate measure of this large middle group.

The factor over which test developers have the most control is the length of the test. Simply put, longer tests will be more reliable than shorter tests, if other conditions are equal. The relationship between test length and test reliability was formulated almost a century ago by Spearman and Brown. The relationship is expressed in Eqn 6.2:

$$r_{kk} = \frac{k\, r_{tt}}{1 + (k-1)\, r_{tt}} \qquad \text{Eqn 6.2}$$

where:
r_{kk} = the reliability of the test k times as long as the original test
r_{tt} = the reliability of the original test
k = the factor by which the length of the test is changed

As the length of a test measure is increased, the chance errors of measurement cancel out and the resulting score comes to depend more and more on the characteristic of the individual being measured, and a more accurate appraisal of the individual is obtained. Of course, how much a measure can be lengthened is constrained by practical considerations. It is limited, for example, by the amount of time available for testing as well as the existence of comparable items to add to the measure. It will also be limited by factors of fatigue and boredom on the part of the individuals being tested. The Spearman-Brown formula presented above can be useful in estimating the amount of increase in reliability that can be achieved by an increase in test length. However, like all prophecy formulas, it is a guide rather than an absolute expectation.

The operations used for estimating the reliability of a measure will influence the estimate obtained. Internal consistency estimates of reliability will yield the highest reliability coefficients because they do not take into account day-to-day variations in the individual nor variations in the individual's speed of work. Test-retest will yield somewhat lower reliability coefficients because they do take these factors into account. Alternate form reliability with an interval between the first and second testings will yield the lowest reliability coefficients because they not only take the above sources of variations into account, but also variations in the specific sample of items presented to the examinee. One needs to be careful in comparing estimates for different measures since differences in reliability coefficients between different measures might arise from the different operations used to estimate reliability. In general, a measure that shows a high reliability coefficient estimated from alternate forms with an interval between testings is the most impressive evidence of reliability one can obtain.

Validity, Reliability and Response Rates

When one undertakes to conduct a survey of a particular population, one attempts to estimate the performance of the members of that population on one or more characteristics. The International Association for the Evaluation of Educational Achievement (IEA) has been conducting such surveys for over thirty years. Studies have been done in the areas of mathematics, science, reading, French and English as foreign languages, computers in education, and preprimary education. In each of these studies, the investigators have attempted to estimate the performance of various age and/or grade populations on various characteristics. The central concern in the conduct of such surveys has been to estimate the performance of a defined population in each nation in particular school subjects. To accomplish this, the investigators have needed valid and reliable measures of the school subject being tested as well as adequate samples of the population being tested. The adequacy of samples is achieved through the way in which the samples are selected (probability sampling) and the size of the samples. A key indicator of the adequacy of the sample being tested is the response rate. The response rate indicates what proportion of the sample originally selected for study actually participated in the testing. Theoretically, any response rate that is less than 100 percent of the selected sample can introduce bias into the results achieved from the testing. However, the amount of bias introduced by a less than 100 percent response rate can never be fully known. Obviously, a response rate of 95 percent is likely to introduce less bias than a sample with a response rate of, say, 40 percent. Investigators strive to achieve as high a response rate as possible, with recent IEA studies reporting response rates of 80 percent or better. All other things being equal, studies with high response rates will yield more accurate estimates of populations than studies that show low response rates.

However, other things are not always equal. Validity is the key consideration in judging the adequacy of a measure. A study that achieves a very high response rate will yield meaningless results if the measures that are used are not valid. It is for this reason that IEA investigators devote considerable time and effort to developing measures that will be valid across a range of nations. Elaborate procedures have been developed to carry out curriculum analyses in each country in a school subject.The results of these analyses are then aggregated and an international set of test specifications developed. These are then reviewed by researchers from the participating nations and a final set of specifications developed. Test questions are then written on the basis of these specifications and sent out to each country for review. The results are used to revise the pool of test questions which may be sent out for another round of review. A final set of test questions is then prepared and tried out in each country. The results of such tryouts are used to further revise the test in terms of the test specifications. The resulting test is usually subjected to a final review by the participating countries before being used. The aim is to insure a test that will possess a high level of content validity across the range of participating countries.

The use of a valid measure on carefully drawn probability samples with high response rates will yield meaningful results. If a test is not valid, then no level of response rate, even 100 percent, will yield meaningful results. On the other hand, the use of a valid test with samples in which response rates are considerably less than 100 percent is problematic. While less than perfect response rates may raise the possibility of sample bias, it is possible that the resulting sample may not be biased. Checks of sample results against population values for selected variables, e.g., gender and income level, may indicate that there is little evidence of bias in the sample despite the less than perfect response rate. In such cases, one can interpret the results with some confidence. However, one would need to point out to the reader the response rate and what evidence one has to judge the adequacy of the sample. If, however, a test is substantially lacking in validity, then even a response rate of 100 percent will not insure meaningful results.

The issue of reliability and response rate is rather more complicated. A test with a somewhat modest reliability, say 0.6, may prove serviceable if estimates of group performance is what is sought. If response rates are high, even a test with modest reliability can be useful. However, if response rates are low and there is evidence of bias in the sample, then study results are questionable. Even if the overall response rate is high, study results may be questionable if the examinees who are not part of the testing differ substantially from the rest of the sample. Thus, for example, exclusion of students in a particular school stream may result in a heavily biased sample and even a test with high validity and reliability will not yield meaningful results. With a high response rate, tests with low levels of reliability will not yield meaningful results at the individual level. For example, correlations between variables at the student level will probably underestimate the true relationships because of the large amount of imprecision in individual scores. Ideally, one would only use a test with a high degree of validity and reliability in studies that achieve high response rates.

Conclusion

This chapter has addressed the issues of validity and reliability of outcome measures used in educational studies. To be sure, the concepts of validity and reliability apply to all measures of educational variables and not just outcome variables. Validity is the key consideration in any measurement operation since it addresses the issue of the meaning of what is being measured. That is, is the variable which is purported to being measured actually being measured. Reliability, in contrast, is concerned with the precision or accuracy of what is being measured. The major forms of validity were identified and discussed. Similarly, the major methods for estimating reliability were presented and compared. The importance of these two key concepts has been heavily stressed throughout the chapter. The reason for this heavy emphasis is that if measures that are used in educational studies are deficient with regard to validity and reliability, the likelihood is that the results obtained through the use of such measures are likely to be worthless.

References

American Educational Research Association, American Psychological Association, & National Council on Measurement in Education (1985). *Standards for educational and psychological testing.* Washington, DC: American Psychological Association.

Feldt, L.S., & Brennan, R.L. (1989). Reliability. In R. Linn (Ed.), *Educational measurement* (pp. 105-146). New York: Macmillan Publishing Co.

Gronlund, N.E., & Linn, R.L. (1990). *Measurement and evaluation in teaching. Sixth edition.* New York: Macmillan Publishing Co.

Messick, S. (1989). Validity. In R. Linn (Ed.), *Educational measurement. Third edition* (pp. 13-103). New York: Macmillan Publishing Co.

OECD (1993). *Education at a glance. Second edition.* Paris: Organization for Economic Cooperation and Development, Center for Educational Research and Innovation.

Rosier, M.J., & Keeves, J.P. (1991). *The IEA study of science I: Science education and curricula in twenty-three countries.* Oxford: Pergamon Press.

Wolf, R.M. (1982). Validity of tests. In H. Mitzel (Ed), *Encyclopedia of educational research. Fifth edition.* (pp. 1991-1998). New York: Free Press.

Wolf, R.M. (1993). Data quality and norms in international studies. *Measurement and Evaluation in Counseling and Development* 26(1), 35-40.

Chapter 7

The Monitoring of Cognitive Outcomes

INA V.S. MULLIS* and EUGENE H. OWEN†

*Educational Testing Service, Princeton, New Jersey, USA
†National Center for Education Statistics, United States Department of
Education, Washington DC, USA

*This chapter describes how decisions are made about the domain of cognitive
outcomes to be assessed, and how this can differ in countries with centralized
systems to those with decentralized systems. The process of developing a
cognitive assessment program is complex, involving numerous decisions about
item format, ancillary materials, and administration procedures. Several of the
innovative alternatives available in designing assessments to monitor student
achievement are described, with some of the benefits of the various approaches
noted. Assessment approaches based on interesting stimulus materials and
procedures for students to construct their own responses are highlighted. Such
innovative assessments promise more meaningful data, although changing the
measures may preclude measuring changes in students' achievement.*

Factors such as history, cultural traditions, values, philosophies of
education, and beliefs about the purposes of education affect the
development of education systems. Since the nature and combinations of
these factors are unique to given national settings, the organization of
education systems differs across countries. These differences can affect
the conclusions about which cognitive outcomes are to be monitored,
who decides what is monitored, how outcomes are monitored, the
purposes of the monitoring, and how the monitoring system is designed.
However, for the purposes of this chapter, the discussion will be confined
to assessments as the major component of the monitoring system and to

OECD countries, with special emphasis on experiences in the United States.

Deciding What Is To Be Monitored

Decisions about the content of assessments should be made in the context of their purpose. For example, assessments and/or examinations can be conducted to regulate students' movement from one level of the education system to another; to certify that individuals have learned a specified amount of information, successfully completed a course of study, or to credential individuals for specific purposes; to collect diagnostic information about students' strengths and weaknesses; or to provide an indicator of the effectiveness of the education system.

Among the OECD countries there are similarities as well as differences in the major purposes of student assessments (Madaus & Kellaghan, 1991; Binkley, Guthrie & Wyatt, 1991). Almost all countries have some sort of exit or entrance examinations or assessments for students planning to move from secondary to university education. Fewer countries have examinations for students moving between primary and secondary education. In Germany, the Netherlands, and to some extent in France, for example, successful completion of the examinations at the end of certain types of secondary schools permit entrance into universities. In other countries, such as the United States and the United Kingdom, separate examination systems are used for admission to university. In the United States, additional examinations are required for admission to post-graduate study and professional schools.

In addition to movement between educational levels, exit examinations can provide certificates or credentials that permit the individual, the educational institutions, the system as such, and employers, to verify that a specific course of study has been successfully completed. These credentials, such as the *baccalauréat* in France, the *Abitur* in Germany, the *Bachillerato* in Spain, the *Diploma* in Italy, the United States and other countries as well as the General Certificate of Education in England and Wales, are given to students who successfully complete the upper secondary school (or its academic stream), and often are obtained after passing an examination or series of examinations. Some systems, including France with the *brévet*, also credential students who complete compulsory education.

Using assessments and examinations at the classroom and school levels to provide diagnostic information for improving individual students' performance has been commonplace for decades. This practice, however, has been expanded in scope to include diagnostic assessments of students and special programs conducted at the state, province or national level. In the United States, some states are using individual assessments for diagnostic purposes, as is France with its National Assessment of 8- and 11-year-olds (LeGuen, 1994).

During the 1990s, as national and local assessments increasingly are being seen as a major component of systemic reform in education, the role of testing to provide information for accountability is gaining in prominence and importance. In the United States, a national council recommended that high performance standards should be set for specific areas of curriculum content, and assessments designed that monitor students' achievement outcomes in light of these standards (US National Council on Education Standards and Testing, 1992). As noted by those engaged in educational reform (O'Day & Smith, 1993) and throughout the various chapters in this volume, however, such a vision has many associated complexities, including the monitoring function itself and those related to the setting of opportunity standards that guarantee that all students have a fair chance of meeting the performance standards.

Who Decides What Is Assessed?

Along with various degrees of centralization among the education systems in the OECD countries and the similarly diverse major purposes of their assessment systems, there also is considerable variability in decision-making about what is assessed. In highly decentralized systems, such as Germany, Switzerland, and the United States, the major responsibility for secondary school exit examinations or assessments falls to the *Länder*, cantons, or states. In the United States, when such tests exist, they generally focus on minimum requirements for high school graduation. The departments or ministries of education in these entities provide for the testing as well as exercise the authority for credentialing based on the examinations or assessments. In the United States, an additional overall national monitoring function is performed by the federal government, although this information does not affect individuals and is used primarily as an indicator of national educational achievement.

In more centralized systems, the national government is primarily responsible for deciding what the assessments will contain, often relying on groups of experts to actually carry out the work of developing the assessments. In some countries, the central government has developed a national curriculum and decisions about what to assess can logically follow the previously made decisions about what is to be taught.

In the United Kingdom, although changes are currently under way, universities traditionally have had a major role in deciding what the examinations would contain, and how they would be used to admit the prospective students to higher education. Currently, at the primary and secondary levels, the country is exploring the introduction of a national curriculum and assessment scheme that would be implemented primarily at the local level, a very ambitious undertaking that will most likely take some time to realize.

Selecting the Assessment Content

Whether or not there is a national curriculum, examinations or assessments in most countries are curriculum based. Even in the United States, where there is no national curriculum, national assessments currently are conducted in six curriculum areas, including reading, writing, mathematics, science, history, and geography (although geography as a separate course is not prevalent in the schools).

When the specific content within the curriculum areas has not been specified through national or state/province level curriculum guidelines, it is generally determined through some type of review process involving educators and policymakers. For example, the content frameworks for the national assessments in the United States are developed through a consensus process involving experts, educators, policymakers, and the business community (National Assessment Governing Board, 1994). Also in the United States, work is being conducted through the national government and professional organizations to develop voluntary national standards that can be used for monitoring at the local and national levels. For example, the National Council of Teachers of Mathematics has published mathematics standards (National Council of Teachers of Mathematics, 1989) and efforts also are proceeding in other subjects, including science, history, the arts, civics, geography, and English.

Monitoring systems generally are designed to assess two aspects of student performance in a curriculum area: a content dimension and a process dimension. The content dimension generally includes knowing and understanding the facts, concepts, and principles that comprise a subject-matter domain. The process dimension covers the skills, processes, and behaviors necessary to use and apply knowledge and understanding about the subject-matter productively. The two content and process dimensions can be used to develop a content-by-skills matrix which serves as a framework for developing an instrument to be used in monitoring cognitive outcomes. Questions or tasks can be developed to provide achievement information pertinent to different cells in the matrix.

As educators and the public have become less willing to be constrained by a rectangular two dimensional matrix, variations on this model have been developed. For example, some international or national assessments have introduced a third or even fourth and fifth dimensions (Robitaille, 1993). These aspects can relate to attitudes or perceptions about the curriculum area, or may be particularly appropriate for only one area (e.g., chronology in history).

With the movement toward educational reform and a growing emphasis on students' capabilities to perform real-world tasks, assessments often need to measure how well students can integrate knowledge and skills across a subject matter domain, or make connections among an array of domains. Many assessment frameworks emphasize such competencies as problem-solving, working in groups, managing resources, and decision-making in an effort to monitor the competencies deemed to be most vital to students' success in later life and work. While some energy has been expended in developing frameworks focusing solely on these processes, this work is still in a preliminary or developmental stage. An example is provided by the report, *Learning a*

Living: A Blueprint for High Performance, prepared by the United States' Department of Labor (Secretary's Commission on Achieving Necessary Skills, 1992).

Selecting Appropriate Assessment Strategies

Once curricular or content frameworks have been established and the areas within the frameworks prioritized, the task of planning the most effective assessment instruments begins. Here, again, a variety of contexts can form parameters constraining the overall approaches to the assessment as well as the formats of the instruments and the questions themselves. Often fiscal considerations will create boundaries for the monitoring effort, as will the breadth of the effort and the time available to accomplish the instrument development. Constructing a broadly based assessment with many questions in a short period of time is very challenging from a number of perspectives, especially if the instruments need to be newly created in their entirety.

In addition to the usual constraints of time and fiscal resources, there also are many technical issues that must be considered in constructing the instruments to measure cognitive outcomes. If new data collection procedures or question formats are to be considered, then decisions must necessarily be made within the constraints of the state of the art of existing measurement methodology. That is, have the procedures been used before? How successful were they and how much additional work will be required to make them operational? Finally, the developers should look ahead to the data collection, scoring, and analysis procedures required to implement the assessment to ensure that these will be feasible. Although having ancillary stimulus materials or equipment may increase the validity of the instruments, such steps may complicate standardizing test administration sessions to the point that reliability is jeopardized. Further, questions that require the students to construct responses, products, or performances may entail very complicated evaluation schemes that require labor intensive scoring procedures. This may affect validity and reliability as well as schedules and budgets.

Item Banking

In beginning to develop assessment instruments, it is possible that questions already exist that are suitable for measuring the established curriculum or content framework. For example, the test development agencies (either government agencies or institutions and companies that develop tests) generally document or "bank" their various test questions using some type of computerized system. The system can contain the text of the questions, as well as an array of information related to each question. Such documentation can include, but not be limited to, coding related to the curricular area and topic being assessed, copyright and data collection information, as well as various statistics from previous field tests or administrations.

Usually, however, existing questions, whether from previous assessments for the same purpose or from other programs or agencies, appear outdated for a number of reasons, including advances in assessment methodology. There also may be issues of security. The availability of questions may mean that they have been used many times in the past. Conversely, it may be difficult to obtain the permission to use some items. Usually the newer and more desirable questions are being used by agencies for their primary purpose and cannot be released for secondary use, because the secondary use might in some way jeopardize the data collected for the initial effort. Also, if new frameworks have been developed, then it is likely that the definition of content areas will be modified, and the questions used in past assessments may not measure the newly defined domain in a valid fashion. Unless a conscious effort is being made to link across assessment forms for some particular purpose—including trying to make direct comparisons of any type, which precludes making changes in the questions—then substantial revisions to existing questions will generally be necessary and occur during the item review process.

Although existing items can provide a valuable source for reviewing previous approaches and stimulating the new instrument development, the above factors tend to reduce the value of existing item banks as sources of questions for measuring new frameworks. It often is best to assume that new instruments will need to be developed in their entirety.

Paper-and-Pencil Assessments: Multiple-Choice vs Constructed Response

In developing assessment instruments, a prerequisite—and crucial—planning decision concerns the question formats that will be used. Will the instrument be designed for computerized scoring? Or, will students be asked to construct their responses orally or in writing? Or, will the instrument employ some combination of approaches?

Computerized scoring provides benefits of resource efficiency. The labor intensive effort of scoring students' answers can be accomplished by machine, thus saving many hours of labor, hence reducing costs and permitting faster schedules. However, to have computerized scoring, the types of answers provided by the students need to be machine readable. Most often, this means using a multiple-choice format whereby students are asked to select among multiple options or alternative answers by marking their choice (usually by filling in an oval or circle). In the most efficient application of this approach, students simply mark their choices to the questions on separate machine-scorable or scannable sheets.

There are, however, other machine-readable question formats besides the traditional approach of presenting a question followed by four or five answer choices, and having students mark these choices on scannable forms. For questions requiring numerical responses, various gridding schemes can be devised whereby the students fill in ovals that relate to corresponding numbers. In a geography assessment, for example, the students can be asked to fill in ovals strategically placed across maps. In a science assessment, students may be asked to draw arrows in a particular place to indicate movement (e.g., wind in a weather question, or blood flow in the circulatory system). Such departures from the regular multiple-choice formats require more sophistication in formatting the cognitive instruments and more complicated scanning procedures. The important consideration is that the instrument developers work with the computer programmers and the individuals who are responsible for printing the instruments according to design formats that can be machine scored.

Despite their efficiency, there is concern about the validity and instructional relevance of instruments based solely on multiple-choice questions (Wiggins, 1993; Office of Technology Assessment, 1992). Because the students are given responses to choose from rather than producing their own answers, such formats do not authentically reflect many real-world tasks. Also, the tasks or problem situations that can be included in such instruments often focus on isolated bits of knowledge, because the need to have one right answer precludes presenting more complex and challenging situations. This can adversely affect instruction, encouraging teachers to emphasize the memorization of—often trivial—facts rather than in-depth understanding of major ideas.

The alternative to multiple-choice formats involves presenting students with open-ended questions or tasks where they are asked to construct their responses. In these formats, students may be asked to provide responses in writing, to provide drawings, to create diagrams, charts, and graphs, or even to build particular models that will either be evaluated by the test administrator or collected for evaluation by trained personnel at a central location. These tasks have the advantage of allowing measurement of a far wider array of outcomes than is possible through multiple-choice testing alone. On the other hand, such questions add expense (because of the need for additional human resources for scoring) and may affect assessment reliability through some variation in administration and scoring. Also, in low-stakes assessments, students may be more likely to omit these tasks, because they require more effort than multiple-choice questions.

Generally, to maximize both efficiency and relevance, it is important to use machine-scorable formats when possible, but also to include some constructed-response formats for more complex cognitive questions and tasks. For example, a mathematics assessment might use multiple-choice formats to assess essential computational understandings, but also include more in-depth problem-solving situations where students would be required to show their work and explain their reasoning. In measuring reading, multiple-choice questions can be used to assess surface understanding and vocabulary, but students also should be asked to demonstrate the depth of their comprehension through the construction of their own written responses. Open-ended questions can be designed that assess students' ability to apply the information contained in texts to formulate ideas and perform practical tasks.

Beyond Paper-and-Pencil Assessment Instruments: Performances,
Classroom Work (Portfolio), and Computers

Although paper-and-pencil instruments are perhaps the most commonly used assessment vehicles, there are other ways to assess cognitive outcomes (Barton & Coley, 1992; Herman, Aschbacher & Winters, 1992). One traditional way is the oral examination or interview. In these instances, the instrument is presented via a trained administrator or interviewer who asks the questions and records students' answers by means of a paper-and-pencil response sheet, or audio or video tape, or some combination of these techniques. The interview technique is appropriate for situations where the student needs to provide an oral response, such as in assessments of speaking and communication skills, music assessments that might require singing, or assessments that might require group-interactive situations. With these types of cognitive instruments, the interviewers can be trained to evaluate students' responses as part of giving the interview or assessment; such evaluations often are part of oral proficiency interviews used in foreign language assessments. However, it often is difficult to standardize such judgments across administrators, so recordings (either audio or visual) of the students' responses can be made and forwarded to a central location for further analysis and scoring.

Full-scale performance assessment techniques may be required to measure specific types of outcomes for some curriculum areas. In these situations, such as in arts assessments, the student may be required to give a performance, which might include playing a musical instrument or performing a dance. In assessments of academic curriculum areas, the students might be required to present their findings from long-term research projects, or design and perform scientific experiments. It should be recognized, however, that such one-on-one assessments of cognitive outcomes are very labor intensive and thus expensive.

Collecting students' classroom work enables assessment of long-term student work, without the expense involved in other data collection methods. Using the classroom work that students have produced as they engage in their instructional programs establishes an immediate connection between the classroom instruction and the cognitive assessment. In the United States, there have been efforts to adapt the portfolio instructional approach for assessment purposes (Gearhart, Herman, Baker & Whittaker, 1992; Gentile, 1992). Since students collect their work as a way of monitoring their progress, the work also can be collected for further evaluation at central locations.

In adapting portfolio instruction to large-scale assessments of cognitive outcomes, two approaches can be taken. In one approach, students and their teachers are asked to select their best work from their classroom portfolios, which can then be collected and forwarded to a central site for evaluation. This procedure maximizes students' ability to submit their best work, but minimizes the possibility of standardizing the assessment procedures. Thus, some portfolio assessments administer several common activities to students. The instrument developers can design activities that will be accomplished by teachers in their classrooms, and then evaluate the students' work according to common criteria. This type of assessment differs from traditional assessments in that the students' work is accomplished as part of their regular classroom activities rather than under standardized assessment conditions. However, the activities and the evaluation criteria can be held constant.

An assessment approach with great future potential involves applying computer technology to assessment. This approach, although not yet used extensively in large-scale applications, involves incorporating the presentation of the instruments with interactive systems that also provide ways for students to record their answers. Computerized assessment also can be adaptive, in that students can be given different sets of questions based on their estimated abilities. Taken to the most efficient extreme, the assessment presentation can also provide for computerized scoring as students enter their responses. Even if the technology does not provide for evaluating the quality of students' responses, it does provide an efficient way to record students' problem-solving approaches so that these can be evaluated subsequent to the assessment.

Interesting Stimulus Materials

Regardless of the approach, assessing students' cognitive outcomes can be made more effective, if the assessment instruments include relevant and interesting stimulus materials. Such materials provide contexts for the assessment tasks and can be a motivating factor. In some instances, these materials can be printed together with the assessment instruments. For example, in a reading assessment, it is important to include authentic representations of actual text materials. This is equally important in other curriculum areas, for example, historical sources can be included in history assessments or scientific journal articles in science assessments. Often diagrams, charts, or maps can be included. Reproducing pictures for inclusion in assessment instruments may often be problematic, but these also can provide excellent springboards for assessment questions and tasks.

Sometimes important stimulus materials cannot be printed in conjunction with the assessment instruments, and alternative approaches need to be employed. For example, the assessment administrator can show an exhibit or series of exhibits to students, or similarly, the administrator can perform a demonstration or experiment. To stand-ardize such presentations, however, it often is preferable to present such stimulus materials via film or videotape. In this way, students all can watch the identical presentation before answering the related questions. For example, the film presentation could be a scene from a play for a literature assessment, a science experiment for a science assessment, or a debate between politicians for a history assessment.

A related solution to providing students with complex stimulus materials is to set up a series of stations where materials are displayed for observation and use by students. Students can carry their assessment booklets containing the questions and spaces for their answers with them as they progress from station to station under a uniform rotation system. This system minimizes the amount of equipment that needs to be purchased, but also limits the number of students that can be assessed in one session to the number of stations involved.

Because stimulus materials sometimes need to provide students with interactive opportunities, or be available for continuous reference or individual use by students, it is sometimes necessary to provide individual students with stimulus materials ancillary to the assessment instrument. For example, students may be given reference materials to work with during the assessment session, including an atlas, a dictionary, a globe, or color pictures—depending on the curricular area and the topics being assessed. They may be given a book of short stories and asked to select one story to read and then answer questions about it. In a mathematics assessment, students can be provided with calculators, rulers, and protractors. The questions asked also might require the use of spinners, manipulative geometric shapes, or other necessary equipment. In these situations, the procedure generally involves arranging for sufficient quantities of materials to be distributed and collected as part of each administration session.

In other instances, the materials cannot be reused across the assessment sessions. The Students may be given a form that actually needs to be completed, or a kit containing some disposable materials to perform a science experiment. In these circumstances, the stimulus materials need to be produced in a quantity that enables each student to have his or her own form to complete or a kit of materials to conduct the experiment. The science kits can be self contained, even including the necessary water, paper towels, or measuring tools. This minimizes difficulties in data collection and helps to ensure standard procedures across sessions. Students actually can create mixtures, connect wires, and sift through materials with sieves and magnets. They can be asked to design the experiment, conduct the experiment, observe the results, and document their findings and conclusions. Subsequent to the assessment sessions, schools can keep the materials that are not disposable, such as magnets and magnifying glasses. The method of monitoring cognitive outcomes also makes it possible for the assessment administrators to help evaluate the degree to which experiments were conducted thoroughly and accurately. Or, students can label the results of their work and place them in containers for subsequent scoring and analysis.

Adopting a Combination of Approaches

In developing instruments to monitor cognitive outcomes, there are many available alternatives regarding the format of the questions, the mode for collecting the data, and the use of ancillary stimulus materials. The approaches described above can be used with individual students or adapted for use in assessing groups of students. Many of these approaches can be used in combination, which often makes for the most efficient assessment and the most useful data in the long run.

Depending on the standards set for students' achievement, the assessment most likely will need to adopt several approaches. For example, an assessment can include instruments for use with individuals and other vehicles that can be used to monitor outcomes for pairs or groups of students working together. For example, a history or civics assessment might need to include measures of individual understanding as well as measures for cooperation and participation. In assessing writing, an assessment might well be much stronger if it included both a standardized component and a classroom-based component. Reading assessments can be strengthened by the use of ancillary materials, and for younger students, reading aloud may be considered the most appropriate approach. In assessing science, it generally is important to have a component that requires students to use equipment to conduct investigations and experiments. Most importantly, assessment developers must design the most accurate and useful measures of the cognitive outcomes possible within the existing constraints.

Constructing the Assessment Instruments

Whether the development effort is managed and implemented by measurement specialists, content area specialists, teachers, or a combination of these experts, several factors must be kept in mind during the development effort. It is important that different perspectives or points of view be taken into account during the development effort by involving various constituents in the development and review process. Also, it is essential that the development effort includes not only the generation of questions to be asked, but also the consideration of stimulus materials, the criteria for evaluating students' responses, and aspects of the data collection procedures. Finally, it is crucial to provide for field testing that encompasses the entire operational sequence, including data collection, scoring, and analysis.

Developing Scoring Criteria

The questions and the scoring criteria for students' responses must necessarily be developed hand-in-hand. If a multiple-choice format is adopted, the correct answers need to be specified and verified as part of the item development effort. Similarly, if students are asked to construct their responses, the levels or degrees of attainment need to be specified and exemplified. This usually entails developing a rationale for partial as well as full accomplishment of the cognitive outcome or standard being assessed. For example, if the student is required to write an essay explaining the causes of a particular war, to design a scientific experiment, or to discuss character motivations in a play, then criteria need to be established describing the characteristics of essays that fulfill the standards, as opposed to those that fall short of meeting the standards. These criteria should be strictly related to the tasks explicitly asked of the students, and not to aspects of performance external to the assignments. Further, it is important that the criteria be specified clearly to provide for a reliable classification of students' work. To assist in training judges or scorers to make reliable classifications, it is customary to illustrate scoring categories with examples. These examples can be hypothetical until field testing, when actual student work is collected. At this point, the scoring criteria will need to be validated by sampling actual student responses and applying the criteria. Training sets, including the scoring rationales, guidelines for the classifications, and examples of student work should be developed to help the judges understand the scoring criteria. This will enable the judges or scorers to practice until they reach agreement and are implementing the evaluation procedures reliably.

Most often, students' assessment responses are collected and sent to central locations to be evaluated. This enables standardized training for the judges and routinized procedures for the ongoing monitoring of scoring reliability. Such procedures can be used for all types of written responses, including drawings and diagrams. Similar procedures also can be used for evaluating and analyzing students' answers and performances captured via audio and video tapes or collected using portfolio methods.

There may be occasions, however, when it is impossible to videotape performances. These types of assessments require that the students' work be documented by trained observers. The observers can record particular aspects of the students' behaviors, and, in some instances, also provide judgments about the quality of the students' work. Also, the students may be required to build bulky models, create certain artwork, or perform science experiments where it is impractical to forward the results to a central location for scoring. These types of assessments require that the students' work be evaluated in the field, using the same types of criteria and procedures that would be used at a central location. Spreading the evaluation process across geographical locations complicates quality control procedures and makes ensuring reliability difficult, but it can be accomplished if necessary.

Defining Data Collection Parameters

Some aspects of the data collection procedures also are inseparable from the instruments. For example, decisions need to be made about the timing for administering the instruments. Many assessment instruments may be administered with standardized timing procedures, whereby all students are given a pre-established amount of time to respond to the assessment. Another issue relates to the amount of assistance that can be given by the test administrators. In most standardized assessments, the administrators are precluded from providing any assistance to the students. However, as more complex performance assessments are being fielded, this is changing somewhat. For example, in being trained to conduct interviews, the administrators may be considered more like facilitators than testors. Although careful limits must be placed on such facilitation, because differential performance by administrators might influence—and possibly contaminate—the reliability and validity of the measures of student performance. Clarity of philosophy and approach is required. Finally, some consideration should be given to whether or not the assessment administrators need any special training. For example, they may need to learn how to operate videotaping equipment or how to demonstrate a scientific experiment.

Prototypes of Ancillary Resources and Equipment

Producing prototypes of any ancillary stimulus materials is also an important part of any development effort. If films, documents, globes, or color pictures are to serve as stimulus materials, it will be impossible for reviewers to evaluate the quality of the assessment questions without seeing the stimulus materials. Additionally, there are operational aspects to consider. For example, if calculators are provided, it is wise to check that they are appropriate in terms of the functions required, the symbols on the keys, durability, and availability. For science kits, it is important to see that the experiments can actually be performed using the materials, and that the materials will conform with safety regulations as well as survive shipping and storage procedures. Finally, inconsistencies in the equipment can affect students' responses and make it difficult to score their answers in a standardized fashion.

Using a Thorough Review Process

Instrument development procedures that ensure a thorough review and revision process can prevent costly mistakes. Such procedures involve including all of the assessment materials in the review steps and making sure that the materials are reviewed from various perspectives. It is very important that all aspects of the instruments be included in the review and revision process. Specifically, the directions to data collectors and students, the stimulus materials, the questions themselves, the procedures for recording or documenting students' answers and/or performances, the scoring criteria that will be used in evaluating their responses, and the way in which the results will be used, must all be assembled and subjected to the review process.

It is equally important that the review occur from various perspectives. The procedures must be clear to all who will implement the assessment, including the administrators, students, and scorers. If the data collectors do not understand the procedures or students do not understand the questions, then the assessment is a wasted effort. Content specialists need to review the instruments for accuracy of information and scoring criteria. They also need to review the complete instruments as well as individual questions or tasks for importance and relevance. Are the instruments really addressing the heart of the standards rather than trivial or tangential aspects of the curriculum? Teachers and curriculum specialists need to ensure that the materials are appropriate for the age or grade level being assessed. Also, measurement specialists need to review the instruments for methodological and operational concerns. Finally, if the assessment will involve students with substantially differing regional or cultural backgrounds and experiences, then the reviews should be conducted by individuals who understand these particular backgrounds. Such reviews can prevent the introduction of test bias that is not relevant to the cognitive outcomes being measured, and may avoid language or situations that might be offensive to particular groups.

Field Testing

The purpose of field testing the assessment instruments is not to collect information about the students' cognitive achievement, but to obtain data about the operational feasibility, accuracy, and reliability of the inst-ruments and the methods for data collection, scoring, and analysis. The purpose of the field test is to "test the test".

Even after detailed reviews and what may seem to be endless and extensive rounds of revisions, it is usually important to have a dress rehearsal before conducting the full-scale assessment. This dress rehearsal can establish how difficult it will be to schedule and conduct the assessments, and provide information from the data collectors about any confusing instructions or procedures. To this end, it is very informative to develop a questionnaire to be completed by the individuals responsible for administering the assessment, which asks about any difficulties they encountered and how the assessment could be improved.

It is also important to gain information from students about how well they understood the directions, questions, and response procedures. To obtain this information, the assessment administrators can conduct interviews after the field-test session. Also, students can circle words they do not understand and write comments or questions about points of clarification on their assessment materials. Accurate information about timing and speededness also can be collected from administrators and students based on how long it takes students to complete the field-test instrument, and adjustments can be made for the final forms.

The field test often presents the first opportunity to implement the scoring criteria with actual student responses. This procedure can reveal additional student misunderstandings in reading the questions. Also, it can reveal unrealistic expectations on the part of instrument developers. In general, it provides an opportunity to clarify, revise, and refine the classification schemes developed to evaluate students' responses.

In analyzing the field-test results, information will be obtained about the difficulty of the questions. Data also can be obtained about the difficulty of the instruments, and how item performance relates to test performance. For example, if large percentages of students who do well on the total instrument but do poorly on a particular question, then the question most likely is flawed. Such item analyses are very useful in detecting ambiguous or unclear questions, or in the multiple-choice format, questions with ambiguous or unclear response options. Also, analyses that relate item performance to overall test performance can be conducted to examine whether particular subgroups of students are having unusual difficulty with specific questions. For example, performance by regional or ethnic subgroups can be examined for differential item functioning (DIF), that is, items that are unduly difficult given the overall performance of the particular subgroup (Holland & Thayer, 1988). These DIF analyses are particularly effective in helping to identify questions where bias has inadvertently been introduced into the instrument by vocabulary that differs across regions or cultural groups, or asking about content that happens to be unique to particular backgrounds. Finally, in cases where final assessments must be assembled to meet statistical criteria for overall difficulty, distribution of difficulty, and mean item discrimination, field-test analyses will allow the test developers to ensure that an assessment has the statistical characteristics optimal to its design and purpose.

Field testing the assessment procedures and analyzing the results will yield considerable valuable information that can be used to improve the instruments prior to the actual assessment. Refinements in data collection methods, more understandable stimulus materials and questions, the elimination of spurious test bias, and increased validity and reliability in scoring criteria are some of the many benefits obtained from a rigorous development process that includes a dress rehearsal of all assessment features.

Measuring Trends in Achievement

That content standards will be revised to meet emerging societal needs is probably inevitable. There also are other reasons why cognitive instruments become outdated and need to be revised to maintain high degrees of validity and reliability. For example, in fields such as mathematics and science, new discoveries are made nearly daily. These scientific and technological discoveries can add to the important content that needs to be covered in the assessments. Or, sometimes, these discoveries can prove that previous thinking was flawed or erroneous, in which case the inaccuracies in the assessment instruments need to be deleted and replaced with current information.

Similarly, evolving research in how children learn can significantly affect assessment procedures. For example, research about effective writing showed this important skill to be the result of an iterative process of planning, drafting, revising, and editing. As such, the ability to write well had little in common with the ability to answer multiple-choice questions measuring isolated aspects of grammar, word choice, and punctuation. To measure how well students could write, therefore, writing assessments evolved to the more valid procedures of giving students realistic assignments, asking them to engage in actual writing, and giving them the time to engage in the process to the best of their ability.

Improvements in assessment methodology also can make existing instruments either outmoded or inefficient. For example, the stimulus materials in reading comprehension assessments were sometimes short pieces written specifically to avoid complexities and ensure correct interpretations. Although more amenable to questions in multiple-choice formats with single right answers, such texts bear little relationship to those that occur in a society, which are replete with ambiguity. Thus, more valid reading assessments must necessarily incorporate authentic texts commonly found in students' daily lives in and out of school. The ability to evaluate students' answers to constructed-response questions using reliable and defensible procedures has made introducing such complexity feasible.

Thus, there are many reasons that instruments need to be updated on a regular basis—the curricular goals will change, the content itself will change through advances in the field, and assessment methodology will improve. This, however, may leave the instrument developer in the predicament of having to meet simultaneous—and conflicting—goals of both measuring trends in educational achievement and providing information based on the most forward-thinking content and methods.

By definition, to measure change it is crucial to not change the measure. To ensure that data from trend assessments reflect changes in student performance and not changes in the assessment instruments and procedures, it is important to maintain many of the questions and hold all the procedures constant from assessment to assessment. If the changes are not major, an item replacement scheme can be used, whereby as many as one-third to one-half the questions may be updated by including matched, but forward-looking replacements in the instrument for the trend assessment. Extreme care must be taken, however, to ensure that the newly developed materials reflect the same aspects of the assessment framework. This procedure provides for stability and linking across assessments based on the comparisons of identical questions, while still introducing innovations and keeping pace with developments in assessment methodology and research about learning in each curriculum area.

If major changes are required to assess the outcomes of a substantially revised curriculum, then dual assessment systems will need to be implemented if the desire is to maintain the trend lines. The instruments and methods from past assessments will still need to be used to establish the links to the past. At the same time, the newly developed assessments can begin to establish new trend lines into the future, with a focus on curricular goals more suitable for the future. When the new trend lines begin to emerge, then the assessment linked directly to the past can be discontinued.

Conclusion

The developed countries have traditionally used a variety of examinations and assessments to monitor students' progression through levels of schooling and certify that certain skills have been mastered or programs completed. With the growing concern about education standards and educational reform, more effort is being expended in assessments that provide information about students' strengths and weaknesses, and that indicate how well the students in countries are doing academically.

Sometimes the assessment frameworks coincide with a national curriculum and in other instances they are being newly created by educators and policymakers to address integrated thinking skills considered important to students' future success. The increasing emphasis on assessment as integral to systemic education reform makes meeting the challenge of providing valid, reliable, and useful assessment data an extremely important endeavor. There are many methodological alternatives available, including open-ended formats, ancillary stimulus materials (i.e., calculators, science kits, and atlases), performance assessments, portfolio or other classroom-based methods, and interactive technology via computers. Used in conjunction with more efficient multiple-choice formats, each can yield valid and reliable data that are meaningful to the users of the assessment results. In designing and developing innovative assessments, it is important to engage in thorough reviews from multiple perspectives and field test both operational and measurement features.

In updating curriculum standards and developing forward-looking assessments, it should be recognized that dramatically changed assessments cannot be used to measure trends to the past. To maintain trends, the implementation of both old and new systems would be required for several years. As the new trends emerge, the assessments linked directly to the past can be discontinued.

References

Barton, P., & Coley, R. (1992). *Assessment sampler*. Princeton, NJ: Educational Testing Service.

Binkley, M.R., Guthrie, J.W., & Wyatt, T.J. (1991). A survey of national assessment and examination practices in OECD countries. Paper presented at the OECD/INES General Assembly, September 1991, Lugano, Switzerland. Paris: OECD, Center for Educational Research and Innovation.

Gearhart, M., Herman, J., Baker, E., & Whittaker, A. (1992). *Writing portfolios at the elementary level: A study of methods for writing assessment*. (CSE Technical Report 337). Los Angeles, CA: Center for Research on Evaluation, Standards, and Student Testing.

Gentile, C. (1992). *Exploring new methods for collecting students' school-based writing: NAEP's 1990 portfolio study*. Washington, DC: National Center for Education Statistics, US Government Printing Office.

Herman, J., Aschbacher, P., & Winters, L. (1992). *A practical guide to alternative assessment*. Washington, DC: Association for Supervision and Curriculum Development.

Holland, P.W., & Thayer, D.T. (1988). Differential item performance and the Mantel-Haenzel procedure. In H. Wainer & H.I. Braun (Eds.), *Test validity*. Hillsdale, NJ: Laurence Erlbaum and Associates.

LeGuen, M. (1994). National assessment in France. In A.C. Tuijnman & N. Bottani (Eds.), *Making education count: Developing and using education indicators*. Paris: OECD, Center for Educational Research and Innovation.

Madaus, G.F., & Kellaghan, T. (1991). *Student examination systems in the European Community: Lessons for the United States.* Washington, DC: Office of Technology Assessment, United States Congress.

O'Day, J.A., & Smith, M.S. (1993). Systemic reform and educational opportunity. In S. Fuhrman (Ed.), *Designing coherent policy: Improving the system.* San Francisco, CA: Jossey-Bass.

Robitaille, D.F. (1993). *Curriculum frameworks for mathematics and science.* TIMSS Monograph No.1. Vancouver: Pacific Educational Press.

US, National Assessment Governing Board (1994). *NAEP reading consensus process. Reading framework for the 1992 and 1994 National Assessment of Educational Progress.* Washington, DC: National Assessment Governing Board, US Department of Education.

US, National Council of Teachers of Mathematics (1989). *Curriculum and evaluation standards for school mathematics.* Reston, VA: National Council of Teachers of Mathematics, Commission on Standards for School Mathematics.

US, National Council on Education Standards and Testing (1992). *Raising standards for American education.* Washington, DC: US Department of Education, US Government Printing Office.

US, Office of Technology Assessment (1992). *Testing in American schools: Asking the right questions.* Washington, DC: Congressional Board of the 102d Congress, US Government Printing Office.

US, Secretary's Commission on Achieving Necessary Skills (1992). *Learning a living: A blueprint for high performance.* US Department of Labor, US Government Printing Office.

Wiggins, G. (1993). *Assessing student performance: Exploring the purpose and limits of testing.* San Francisco, CA: Jossey-Bass.

Chapter 8

The Monitoring of Affective Outcomes

JUDITH TORNEY-PURTA

University of Maryland at College Park, USA

The monitoring of affective outcomes encompasses quite a different set of issues than those connected with cognitive outcomes. This area has, in some respects, made enormous progress in the 1980s and early 1990s; in other ways the field is much the same as it was in 1980. A number of new methodologies have been proposed, some from research on attitude and opinion measurement in adults. Efforts have been made to incorporate models from the rapidly developing fields of cognitive and social psychology into attitude measurement, blurring the line separating affective outcomes from other types of outcomes. Attempts to measure, monitor, and set standards for the cognitive achievements of a country's pupils remain much more common than attempts to do so for affective outcomes, in large part because it is very difficult to obtain agreement about the appropriate role of the school in the development of values and attitudes (and hence about their appropriateness as outcomes). Three models for assessing affective outcomes are examined: the psychometric model; the criterion referenced model; and the social information processing model. The domain of civic education is given special attention. Also addressed are some complex issues going beyond the methodology for assessing affective outcomes and setting standards.

While thinking about how to assess and monitor outcomes, and how to set standards for them has advanced rapidly in the cognitive domain, it has also been recognized that most schools also have aims beyond those associated with the fostering of cognitive achievement. Other, innovative ways of thinking about educational models and a richer mix of methodologies for data collection than the one currently characterizing most monitoring of educational progress have also been called for. These calls appear in studies sponsored by nearly every international organization with a responsibility in the education sector, for example, the Council of Europe, OECD, and UNESCO (e.g., see Westin, 1989; Taylor, 1993; Peschar, 1993; UNESCO, 1992). They also appear in the publications of national groups with an interest in the monitoring of education progress. Examples from the United States are the Board on International and Comparative Studies in Education (BICSE, 1993) and the Special Study Panel on Education Indicators (1991).

The problems raised by affective outcome indicators range from relatively technical and methodological ones—especially the construction of measures and how to monitor them over time—to much more subtle questions about the meaning of freedom to hold one's own attitudes or beliefs in a democratic society, and the freedom of the family to shape children's views. This chapter will raise some of these issues for debate. It will also suggest a new model for conceptualizing assessments used for the monitoring of educational progress.

Affective Outcomes Across Nations

Two Domains — Self and Society

The realm of affective outcomes is very extensive. This chapter will focus on only two relatively distinct and limited domains of affective outcomes. The first includes those related to the *individual student* and that young person's motivation to do well in school as a preparation for the realization of potential in later life. The second domain is related to the *community and the society* within which the student lives, the group's need for harmonious social relations, and its wish to provide continuity through the transmission of certain traditions or views about social or political institutions. In some societies undergoing rapid economic and technological development these two domains merge, and the individual is exhorted to learn in order to serve the needs of society, for example, achieving in mathematics or science because the country needs skilled workers in these fields. In North America and most European countries, it is somewhat easier to separate the domain of individual or self from that of group or society.

Even within these two domains, however, it is difficult to draw a precise line between the cognitive and affective outcomes of schooling. An affective outcome is much more likely than a cognitive outcome to include an evaluative or feeling dimension and to be related to a behavior, but cognitive representations are also present. The next section presents one illustration from the individual or self domain and two from the societal or group domain.

Three Illustrations

Current views of individual motivation in school focus on cognitions, feelings, and achievement-related behaviors. On the cognitive side are attributes or explanations for success or failure related to factors such as effort, ability, or luck. Examples are statements such as "I failed that math test because I did not study hard enough" (an effort attribution) or "I failed that math test because I am not smart in math" (an ability attribution).

Research studies have found substantial differences between countries—notably the US, China and Japan—in these cognitions or attributions about the role of effort and ability (cf. Burstein & Hawkins, 1992; Stevenson & Stigler, 1992). Pupils are asked why they believe they have succeeded or failed in specific situations; in other words, to what they attribute success or failure. This type of cognition about the self and the environment can be seen as an affective outcome, but it presents very different issues of measurement compared with those faced in assessing cognitive outcomes related to the learning of specific subject matter, for example, the understanding of a mathematical problem or knowledge of a certain historical incident. The assumption is that the reported attributions are related to behavior—the student who believes that effort rather than ability or luck is essential for achievement being more likely to behave by studying hard and attending carefully to the demands of the teacher.

There is also a feeling or affective side to this self-motivation domain. This is often referred to, contrastingly, as a need to achieve or a fear of failure, a high or a low value placed on achievement in a particular subject area (or in general), a feeling of self competence or incompetence, or a feeling of self-efficacy or inefficacy in achieving chosen or valued goals. Schools in most countries attempt to enhance both aspects of motivation—the cognitions or attributions about the importance of effort in achievement and the personal sense of competence and interest in subject matter. Thus measures or assessments of both types of characteristics are important. Problems arise in cross-national attempts to measure these aspects of self concept that are to student goals or academic achievement because of difficulties in translation, both in linguisitic and cultural equivalence. In some countries, for example, the notion that an individual can select self goals that are separate from those enunciated by the family is not readily accepted.

Moving to the second domain, that of society, there is a recognized need in most nations to promote positive relations between different ethnic, religious, racial, or linguistic groups and some sense of social cohesion in order to be able to achieve valued goals for the society as a whole. Until the early 1980s most educators focused on the affective and behavioral aspects of these intergroup relations, for example, a student's negative self-evaluation or prejudiced feeling about those from other racial or religious groups, and associated behaviors of discrimination, or hostility. In the 1980s and early 1990s pychologists have noted the extent to which the cognitive images or representations of members of the group in question are influential in intergroup relations and are related to the feeling and behavior dimensions. Avery (1992) found that affectively intolerant students were especially likely to possess negative images or cognitive representations of disliked groups, and they were also likely to predict that granting these groups rights to meet or protest would have negative consequences, often violent confrontation. Kuklinski and colleagues (1991) also found that cognitive arguments supporting generalized tolerance were often interferred with by negative emotional responses to groups such as the Nazis. Likewise, social identity theory is built on the premise that young people maintain a positive cognition of their own membership groups in order to enhance their own feelings of self worth. They behave toward out-groups in negative ways to further enhance their sense of belonging to their own valued group. It is not easy to say where the affect ends and the cognition begins when discussing intergroup relations. As in the self motivation domain considered above, the assessment of cognitive representations of other groups in society requires different tools than the assessment of cognitive representations of a chemical reaction or of grammatical sentence structure. Assessing the feeling or behavior associated with these representations presents still further challenges.

Another important issue in the domain related to the demands of a society is the expressed need to pass on to the future generations certain public knowledge and public values, attitudes and group identifications. This process, related to the civic education aims of the school, includes passing on knowledge of the history and structure of political institutions, but it also includes inculcating feelings of loyalty to the nation and its symbols as well as encouraging belief in fundamental values such as the rule of law. Most democratic societies also wish to foster certain behaviors in students, for example the skills necessary to participate in politics by voting, analyzing issues, or cooperating with other community members toward a common goal. Often behaving in a way that recognizes the equality of those from different ethnic, national, racial or linguistic groups (positive intergroup relations) becomes a part of civic education as well. The research related to this area is often conducted under the rubric of political socialization.

Cha, Wong and Meyer (1988), in their analysis of course offerings in more than 60 countries over a period from 1920 through the 1980s, note that education in this societal domain is offered under different course titles in different countries at different time periods—civic education or social studies in one, moral or religious education in another. Moreover, in the area of civic education the cognitive, feeling, and behavioral dimensions are not easy to separate. And there is considerable argument about how much history or how many details of governmental or constitutional structure a pupil should know in order to become a well functioning citizen in a democracy.

In the United States attempts are being made to define a set of voluntary national content and performance standards in the area of civic education, and distinctions are being made among a number of aspects of importance for citizenship in a constitutional democracy, e.g., knowledge of history as well as of the political and economic system; skills in thinking critically about public issues and working cooperatively with others. Also described are traits of civic character, such as feeling a sense of responsibility for oneself and for others, behaving in ways that show tolerance for individuals from diverse backgrounds, showing an internalization of principles such as freedom of speech, and feeling potentially effective in seeking to change negative situations in the community (Center for Civic Education, 1993). This project is also seeking to develop indicators and assessments of these wide ranging outcomes of civic education, in many cases the student's ability to analyze or explain a situation where a democratic principle applies.

An effort by the OECD to define indicators in the areas of political knowledge and values cannot be placed neatly into the cognitive, affective, or behavioral realm either (Peschar, 1993). OECD's project on indicators of education systems (INES; see Chapter 3) seeks to develop measures of what are called Cross Curricular Competencies (CCCs)—indicators of the students' readiness to meet life challenges, including self esteem, communication skills, problem solving abilities, and political understanding. It is part of a larger OECD-led effort to define education indicators and to monitor the standards of education on a regular basis across a variety of subject areas (OECD, 1993). Issues similar to those raised in the US standards setting project were identified in the Netherlands and the United Kingdom as part of the preparatory work of the CCC project. Thus in the area of civic education as well as in the domains of intergroup relations and achievement motivation, there has been a movement toward simultaneously considering ways to assess cognitions about the self, the educational system, the community, the society and the political system, along with feelings and behaviors related to these areas and issues. These are issues to which current models cannot easily be applied from the content or performance areas where indicators and standards have been more completely developed.

In summary, two domains of interest in the affective outcomes of schooling have been identified. The first pertains primarily to the individual and his or her sense of competence and motivation to achieve. The second is defined in terms of a society's needs for societal cohesion within pluralistic constitutional democracies and for certain aspects of civic character including feelings of respect for the rule of law and behaviors demonstrating tolerance between groups.

Indicators in the Self and Societal Domains

There are major difficulties in identifying and measuring indicators in the two domains that will be useful in educational monitoring. First, the valued outcomes of motivated students, tolerant students, and students who support democratic government are influenced as much by out-of-school as within school factors—by the familes from which students come and to which they return each evening, and by their peers whose opinion is essential to them both in and out of school. These outcomes are also related to the political situation in the country or local community. For example, a newly democratized state experiencing rapid economic and political change requiring sacrifice from its citizens and providing dim hopes for youth employment in the near future presents a very problematic context for civic education. The power of television and video to shape students' awareness of social and political issues is also a factor. MTV and CNN have created a world-wide video culture that has a substantial impact on attitudes and behavior.

A second difficulty, even if one focuses on the school's influence on affective outcomes, is the difficulty of pinpointing any one or two subject matter areas in which motivation and sense of competence, tolerant inter-group behavior, and civic or moral character or national loyalty are meant to be fostered. It is important to have students who feel at least minimally competent and interested in all subjects, and who believe in the importance of effort to achievement in general. One addresses intergroup tolerance and fosters the internationalization of democratic values and national loyalty in social studies, history, and related subjects, but also by reading national literature in the mother tongue and through performing national rituals, as well as in the climate of relationships between teachers and pupils and between classmates throughout the school day.

This difficulty is reflected in the heading "cross-curricular comp-etencies", which is used as the name for a study conducted as part of the OECD indicators project (Peschar, 1993). This study has also noted the diversity of ways in which the OECD countries represent their goals in various syllabi and the very small number of countries which attempt to assess the "noncognitive" goals beyond individual teachers' assessments on reports to parents.

A third problem with this diverse but important set of issues is that the individual student and his or her own construction or processing of the information shapes what results from the school's efforts. For civic education or intergroup education to be effective means that the students must give personal or private meaning to the values, group identifications and messages of the school, and internalize them as guides for their own behavior. The representations of civic society or of ethnic groups or of the nation's history that individuals internalize, along with the associated feelings, often differ substantially from those that the schools have attempted to foster because of factors such as individual differences in the processing of information, experience which has led to the wish to challenge authority and the status quo, or strong group identifications or beliefs encouraged by the family or community. For example, countries all over the world find themselves with adolescents who are alienated from the political system, particularly young people who are about to leave school and believe they are unlikely to find employment. They may understand lectures about the history of groups which achieved economic well-being through hard work, the value of equality for immigrant workers, or the relationship between politics and the economy in a way which differs substantially from the intended message of the school.

Importance of Affective Indicators

All this said, however, constructing indicators for the cognitions, feelings, and behaviors associated with affective outcomes remains a challenge which cannot be ignored in a period of rapid political and economic change, increasing migrations across borders, and often violent confrontations between alienated youth and those from minority or immigrant groups. Countries as diverse as Germany, Sweden, the United Kingdom and the Republics of the former Soviet Union share a concern about the ways students from different cultural, national, linguistic and racial groups get along and about the possibilities for violent outbreaks. For examples, see Harber (1991) on anti-racist education in the United Kingdom, and Iudin (1993) concerning students in the former Soviet republics and their attitudes toward those of other nationalities, as well as reports of international organizations such as the Council of Europe calling for action to prepare European students for life in democratic and pluralistic societies. Rapidly democratizing countries in all major world areas are seeking guidance not only about how to go about reforming their curricula to prepare students for these new political and economic realities but also how to judge their success or failure in doing so. A number of scholars in countries such as the Czech and Slovak Republics and Poland have recently conducted youth surveys, in some cases repeating questions regarding political authorities and the meaning of concepts such as democracy (Fatygi & Szymanczaka, 1992).

In 1991, a multi-national group of researchers conducted a survey comprising 3250 young people ranging from 13 to 17 years of age from eleven European countries (Bulgaria, Rumania, Hungary, the Czech Republic, Poland, Russia, Finland, Norway, France, Germany, and Switzerland) and the United States (Alsaker & Flammer, 1994). This EURONET project includes a special focus on subjective sense of well-being, daily activities including time use, significant life events, beliefs about personal control and adolescents' future orientations. The concerns of the researchers include finding predictors of well-being and specifying how the cultural context has an impact. This study seems to merge concerns for perceptions of self competence with views of the societal and political future (see also extensive work done by one of the EURONET collaborators on views of the future; Nurmi, 1991). This project has not explicitly been designed as an indictors effort, and neither standard-setting nor monitoring over time seems to be envisioned.

A report sponsored by UNESCO and conducted by the National Foundation for Educational Research in England and Wales, surveying policymakers and curriculum experts in eight countries of Eastern Europe and 18 countries of Western Europe reported on the rankings given to 10 types of values relating to education objectives (Taylor, 1993). Eastern Europeans gave precedence to citizenship, democracy, and national consciousness; Western Europeans gave their highest rating to intercultural education, international understanding, and anti-racism. These are precisely the topics associated with affective outcomes relating to society outlined above, although it is important to be aware that responses were obtained from a relatively small group of "stakeholders" in the societies surveyed. The report further noted that these themes often must be handled spontaneously by the teacher as a situation arises, and are very seldom incorporated into outcome indicators.

The remainder of the chapter will concentrate on the domain relating to society, with examples chosen particularly from civic education, with some references to moral education and intercultural education.

Two Measurement Models for Affective Outcomes

The two approaches to measurement in the domain of society which introduce this section are relatively long-standing in the literature. First is the psychometric approach using Likert scales of attitudes, often adapted from studies of adults, for example, measuring constructs such as the sense of political efficacy or the leaning toward authoritarianism. With secondary school students these are often quite useful, although long items or those including complex phraseologies may confound attitudes with measures of reading ability or word knowledge. With students at all levels, Likert scales often present problems of translation to ensure cross cultural equivalence (Brislin, 1986); for example, dichotomies such as "internal" and "external" control may lack the necessary nuance to understand motivation in other countries (as Burstein and Hawkins, 1992, indicate). Problems of socially desirable respondse and trait desirability may also create difficulties. The discussion of the IEA Civic Education project later in this section describes some ways of using extensive piloting, careful back translation, scale analysis, and multiple measures to deal with some of these problems.

It is difficult to conceive of the establishment of standards when the usual 5- or 7-point agree-disagree Likert-scale format is used with attitudinal statements. The usual methodology is to choose items which vary in their extremeness, and then use through factor analysis or some other way of assessing whether a unidimensional and reliable scale exists. Such approaches also seek to maximize meaningful variation between individuals, by ranging them, for example, along a dimension from high level of support for civil liberties to low level of support. Consider the following hypothetical problem. If a four-item attitudinal scale has a range from four to 20, with the respondent who scores 20 being the one who has chosen "strongly agree" for every item supporting civil liberties and "strongly disagree" for every item negative to civil liberties, it is very difficult to set a standard before (or even after) the data are collected without violating the principles of scale construction. Is the only response pattern that meets the standard to be a 20, and consequently, is an individual who only agrees or disagrees but not strongly (and thus obtains a score of 16 on the scale) failing to meet the standard? Or is another who selects the mid-point of uncertainly on three of the four items because she is balancing individual rights against a feeling that there are some areas where the state should intervene against dissidents (and thus obtains a combined score of only 14) failing to meet the standard? Further complicating this issue is that a very high score on some attitude scales may be almost as problematic as a very low score. Moreover, when attitudinal measures are administered there is often an explicit statement that these are not items with right and wrong answers, but that one should freely give an opinion. If students knew they or their schools would be judged on their answers, there would be a strong pressure to give a socially desirable response. Although it would in theory be possible to use attitudinal measures as cross-time indicators, issues such as trait desirability make the setting of standards extremely difficult. This issue is discussed further in the concluding section of this chapter.

A second model, not wholly distinct from the first, is exemplified in assessment approaches which have strong ties to national curricular objectives, and sometimes even to specific courses. Something like criterion referencing could be thought of for attitudinal and affective as well as cognitive measures. The clearest illustrations (though they are not perfect examples of criterion referencing) are courses in moral education. These can be found especially in Asia but also in some European countries with a strong emphasis on religious education.

In Taiwan, for example, moral education permeates the entire school curriculum but especially the courses labeled "civics", "morality" and "Chinese Language and Literature". Many of the stories included in the literature to be studied have explicit morals, and the students' ability to grasp these lessons is part of the evaluation. A grade for moral conduct also appears in each student's progress report. There is apparently some attempt at standardizing teachers' use of the rating scale by separating the ratings for specific moral themes such as "public spirit", a deep appreciation for others' help, and unremitting efforts to improve oneself. An assigned teacher for each class reviews these ratings and follows up with parents when students' moral behaviors appear to need strengthening (Lee, 1990). The Confucian tradition is often the basis for the list of eight moral virtues, four cardinal virtues, three pervasive virtues, and five eternal virtues to be inculcated and assessed. In one sense may be considered as standards.

Moral education courses in Singapore are driven by a national goal of strong government with economic growth and social stability. Assessments have been more proposed than implemented, and usually involve questions relating to moral issues either in examinations in language or religious knowledge or in tests for courses in moral education (which is not an examination subject). Gopinathan (1988) in his discussion of these issues comments on the difficulty of separating the cognitive from the affective and behavioral aspects of this subject.

Examining the first survey of civic education undertaken by the International Association for the Evaluation of Educational Achievement (IEA), provides some illustrations of the psychometric approach and, to a lesser extent, the criterion-referenced model for affective outcome measurement and analysis.

The IEA Civic Education Study

The IEA conducted a survey in civic education relying on the type of Likert scales described above as well as on certain criterion-referenced measures and other scales eliciting representations of various political institutions. The surveys elicited responses in 1971 by 10- and 14-year-olds and by students in the last year of upper secondary education in the Federal Republic of Germany, Finland, Ireland, Israel, Italy, the Netherlands, New Zealand, Sweden, and the United States. A total of 30,000 students responded to these instruments; more than 5000 teachers replied concerning pedagogical practices, and 1300 principals and headmasters described the schools (Torney, Oppenheim & Farnen, 1975; Torney-Purta and Schwille, 1986).

This was the earliest IEA study to place such importance on attitudinal scales, with some attention given to behavior as well. Nearly half of the instrument was made up of attitudinal measures, many of them adapted from previous studies of political socialization or from studies of adult political opinion. In a rating instrument entitled "how society works", students described their views of social institutions such as "political parties" and the "police". The cognitive test was based on an analysis of the curricula of some participating countries. A number of questions also elicited information about the students' concepts of democracy, and about economic processes such as tariffs. All attitudinal and cognitive measures were pilot-tested extensively; back translations were carefully checked; and a factor analysis of the scales was conducted to ensure psychometric similarity in the different countries.

The Member countries of the IEA consortium of research institutions select in which studies they wish to participate. When the decisions were made—in the late—1960s for the Civic Education Study only Western industrialized democracies chose to take part and completed the study with full data. That makes it even more striking that the between-nation comparisons on attitude scales indicated that none of these participating countries "had a uniformly high level of success in transmitting civic values, perhaps because subtle incompatibilities exist" (Torney-Purta & Schwille, 1986, p. 34). In no country did the students score above the international mean on all three of the attitudinal/behavioral factors. Specifically, among 14-year-olds, the countries in which average support for democratic values was above the mean for all countries fell below the mean for all countries in average support for national government and civic interest/participation. In some respects this examination of patterns of attitudinal responses across countries might be thought of as a first step in *a posteriori* standard setting.

Regression analyses were then conducted for three criterion variables: cognitive civics achievement, anti-authoritarian attitudes, and participation in political discussion (Torney et al., 1975). The socioeconomic status of the home was a moderately powerful predictor of 14-year-olds' scores on the cognitive test of civics and of anti-authoritarian attitudes in all countries. In contrast, socioeconomic background was not a statistically significant predictor of participation in political discussion for 14-year-olds in any country. Coming from a home of high social and educational level was associated with experiences which influence knowledge and basic democratic values but not with motivation to be an active participant.

One of IEA's most important purposes has been to relate aspects of classroom practice or educational policy to levels of outcomes achieved in social studies or civic education. Home background, age, sex, and type of school were controlled statistically in order to assess the extent to which learning conditions were significant predictors of civic education outcomes (Torney et al., 1975; Schwille, 1975). The findings were remarkably similar cross-nationally. In all countries the extent to which students reported that teachers encouraged the expressions of opinion in the classroom—a measure of classroom climate—was related to high scores on the cognitive tests and to self-reports of participation in political discussions among 14-year-olds. A high degree of openness in classroom climate was also related to less authoritarian attitudes for both 14-year-olds and pre-university students.

Students who reported the frequent practice of patriotic rituals in their schools (ceremonies with the flag or singing patriotic songs) tended to be less knowledgeable and more authoritarian than students who spent less time on patriotic observances. The amount of class time spent on printed drill or on memorizing facts and dates also predicted lower scores on the cognitive test and more authoritarian attitudes in some countries.

In short, a reported stress on rote learning and on patriotic ritual within the classroom tended to be negatively related to civic education outcomes, while the opportunity to express an opinion in class had a positive relationship. Of course, one cannot infer cause and effect from these correlational analyses. Since the socioeconomic status of the family and the type of school (academic or vocational) were controlled statistically in regression analysis, these findings have somewhat more weight. Further, these findings were remarkably similar across nine OECD-like countries with different educational and political contexts.

Some of the scales from IEA's civic education study could have formed useful indicators if they had been repeated in the period from 1971 onwards. Tracking at 5 to 10 years intervals the level of support for equality or for women's rights would have provided useful information. If and when the study is repeated, it is probable that some of the findings will remain similar, for example, the positive value of a classroom climate in which students' opinions are respected and the negative contribution of rote and ritual.

Challenges in Indicator Development

Societies continually face new realities as well as areas where long-standing problems intensify. The range of countries interested in civic education contributing to democratic values has increased enormously since the late 1980s and early 1990s. There have been strong (and sometimes aborted) movements toward democratic reform in every major world region. What are appropriate indicators of success for schools which for the first time are faced with preparing students to participate in competitive elections or to read accounts of government policy in independent media? This is a special challenge because both teachers and parents in these emerging democracies received a very different kind of civic education than that which they are being asked to provide and evaluate. Is it reasonable to expect that students will be optimistic rather than cynical, or that they will be firm supporters of abstract ideals of democratic government? How can one construct measures when even the definitions of terms like democracy and participation have changed so much from what was taught a short time before?

Resurgent authoritarianism and racism presents an urgent need to better understand intergroup relations. A number of items dealing with authoritarianism, tolerance, and support for equality for all groups were included by the IEA in the instrument used in 1971, and many of these scales were strong enough in their reliability and validity that they could be used to form indicators. This study, along with many others, found that most students verbally endorsed tolerant views. Behavioral indices suggest a much less optimistic picture, and the measurement problem for indicators and standards is a serious one here, especially the problem of students' tendency to give what they know to be the socially acceptable response.

There are important potential process indicators in the affective area as well. The first IEA study of civic education found clear evidence that students who participated in classroom discussion of issues rather than memorizing dates or facts about politics or participating in patriotic rituals had both lower scores on the authoritarianism measure and higher scores on the knowledge test. There was, however, insufficient data across countries to provide fully satisfying indicators of an implicitly democratic curriculum. This factor may be of such importance that countries should establish as an important indicator the extent to which teachers are prepared and supported to create a democratic climate in their classrooms. In Eastern Europe and the former Soviet Union, where teachers who were trained and taught in an authoritarian system may be wary of discussions of citizenship or politics, this training and support may be vital.

Many countries have experienced considerable changes in attitudes toward women's political rights and in the number of women serving in high political positions since the early 1970s. IEA's first civic education instrument included four short items measuring support for women's rights. Even these items showed interesting between-country variation and enormous gender differences across countries—with females much more supportive of women's rights than males. A widely applicable and appropriate cross-national measure of what women's rights and equality mean to students is badly needed, and could serve as the basis for long term monitoring. Reading the curricular objectives of some countries suggests that one might even be able to gain agreement on a standard in this area in some regions of the world. It may also be plausible to consider as an indicator the size of the gender differences in important aspects of political knowledge and participation—as gender differences in reading achievement have been included in *Education at a Glance II* (OECD, 1993).

The first IEA study included no items on the environment, because it was not yet on the public agenda. Issues relating to the ecological health of the planet and its future are much more likely to be part of the everyday awareness and concern of youth than issues such as political parties. Many students engage in recycling behavior or are trying to encourage adults to express more concern for the environment. This is often their introduction to political participation and to basic aspects of political awareness, such as the idea that group action can often be effective when individual action is not. There have been studies of environmental attitudes conducted cross-nationally, but the items have often suffered from all the deficits of Likert scales, being especially complex and difficult to translate. In principle, however, this is an excellent content area to begin to think about indicators and standards. It is well suited to links between attitudes and behavior; it matters to students and mobilizes their interest in answering questions; it can be phrased in a problem solving format which links cognitive and affective elements; and it is well suited to analysis within the social information processing framework, to be discussed in the next section.

The Third Measurement Model for Affective Outcomes

In addition to the psychometric model and the criterion referencing of affective outcomes to syllabi in areas such as moral and religious education, the social information processing model that has emerged in the fields of developmental and social psychology has promise for guiding assessment and monitoring in the affective domain.

This model has arisen since the late 1970s and builds on assumptions about the social and personal construction of knowledge by students in which cognition or reasoning, feelings and intended behaviors all have a given place. There are several "social information processing models"— primarily proposed by cognitive and developmental psychologists. There is one for understanding moral reasoning and behavior (Rest, 1983); one for understanding healthy peer relations (Dodge, 1986); and another one for solving or making decisions about social problems (Voss, Tyler & Yengo, 1982; Torney-Purta, 1992). This chapter marks the first time that these models have been considered in a way to highlight their commonalities and to apply them to educational measurement in the affective domain.

The three approaches have in common an attempt to lay out a process comprised of several steps in which the individual engages when confronting a specific situation or problem. At the first step incoming social cues are attended to; at the second step these cues are interpreted and integrated with material stored in memory about previous similar situations. In both of these phases the individual is interpreting the situation as presenting certain opportunities or constraints, as threatening or unthreatening, or as one in which helping another person is or is not called for. In a third phase the individual identifies the alternative responses; at a fourth a response is chosen, and after the assessment of consequences of different responses, the individual responds. For example, programs designed to improve peer relations among diverse groups or to foster socially supportive behavior often focus on helping individuals to perceive the situation in a non-threatening way and to think of alternatives to aggressive behaviors. In moral education programs, the individual is encouraged to perceive certain situations as involving moral principles and to reason about alternative responses taking that into account.

These approaches make one point that is important for assessment and monitoring. It is insufficient to concentrate on any attitude or piece of behavior isolated from the individual's construction of ideas about the situation. To use an example from civic education, it is not simply a "habit of civic participation" that one wishes to foster through democratic socialization. Rather one hopes that students will possess an image of their local government as responsive to citizen action, will understand that group action is generally more effective than individual action, and will perceive problems in their communities as amenable to cooperative solutions and as part of their civic duty (steps 1 and 2); will be able think of alternative ways in which they might work with others to solve these problems (step 3); and finally will choose to try to interest others in working together toward solutions. From the point of view of assessment this means that single item or short scale attitude indicators are unlikely to be adequate. In the ideal situation, the students would be asked to deal with some sort of hypothetical situation in which they are asked to recognize a problem and then to think of alternative ways to solve it. Preferably the problem should be in an area which matters to the young person, for example, the global environment or issues important to their own future.

Recent work in attitudinal and survey methodology is complementary to this social information processing model. For example, Saris and his colleagues in the Netherlands (Gallhofer, Saris & Schellenkens, 1988; Saris, 1991) have developed a model in which it is possible to use computers, computer networks or multimedia devices which present hypothetical vignettes or real information about political issues for respondents' reactions. Some of the attempts to build performance assessments in history in which students are given source materials which they are asked to interpret are similar, as are techniques for problem solving of international political problems administered either orally or in writing (Torney-Purta, 1992).

Another important perspective related to this model is that beliefs, attitudes, and motivations do not exist as stable unitary characteristics within the individual which can be "measured", but are influenced in their expression by a variety of aspects of the social context—the reasons given for an assessment, the order of questions in a survey, the presumed responses of peers, the concerns which matter most to the individual at the time the instrument or interview is completed, and so on (Schwarz & Sudman, 1992).

The implications of this model for monitoring affective outcomes are yet to be explored, but hold promise for more satisfying and justifiable measures and processes of inference.

Conclusion

Monitoring outcomes and setting standards for schools is technically difficult in any domain, but the problems in the affective domain are much more complex than in the cognitive domain because another set of judgments is involved. An evaluation in which the schools are to be judged as effective based on standards involves certain other dilemmas. The remainder of this section will raise for debate a number of positions along a continuum of important issues relating to the setting of standards in the affective domain.

One end of this continuum would ignore the potential complications and take the position that standard setting in the affective domain can operate on the basis of the same principles and using very similar procedures as those used in setting standards in the cognitive domain, constrained only by technical or methodological considerations.

The opposing position on the continuum is that no standard setting relating to affective outcome indicators should take place because of ethical problems assocaited with the respect for the diversity of individual feelings, attitudes and opinions within democratic societies and the limitations on the school in discussing values. According to this second position, when measures such as those discussed above are used, the results should be presented in a way that does not imply any standards. One could report in the form usual in research reports, by comparing groups (e.g., attitudes of students who have studied a special curriculum on alternatives to violence in solving conflicts, compared with others who have not). Or one might report indicators over time in a descriptive fashion without any implied standard (e.g., the mean level of support for the importance of voting on a 10-point scale was 7.8 in 1975 and 8.0 in 1980, and so on). This position concerning standards would not justify any judgments of schools or other educational units as meeting or not meeting a standard based on student responses to attitude scales or reports of past or future behavior.

The intermediate positions on this continuum regarding standard setting suggest a number of constraints. All of these positions might be debated with respect to the conditions under which standard setting in the affective domain would be justified.

One intermediate position is that affective standards should be set only in a content area about which there is such a preponderance of com-munity sentiment that it can be argued that the school is justified in having zero-tolerance for certain attitudes, for example, those which support violence against teachers or other students.

A second intermediate position is that affective standards are appropriate only when there is a high level of community support. Parents and community leaders can be polled regarding the desirability of the content of a "short list" of shared values, for example, honesty and respect for others, and regarding the ages at which the students are expected to achieve these standards. This position becomes much more difficult in culturally pluralistic and democratic societies where the groups holding values different from the majority may be in low status positions—which makes it difficult to have their voices heard. Another way of phrasing this position is to say that affective standards can be set only partially by educational or political leaders, must be extremely sensitive to family and community contexts, and must concentrate on a very short list of affective outcomes about which there is widespread agreement concerning their importance. How approximation to these standards is to be measured is an important but somewhat separate question. International and comparative research could be of considerable value in debating this position, especially by showing examples of success in promoting these standards in pluralistic and democratic situations and ways in which different stakeholders in a society might be involved in discussions of value standards. It can be noted that the OECD is undertaking a 12-country survey of public expectations and attitudes to education, and that the results may be useful in benchmarking some of these concerns.

The third intermediate position on the continuum is that affective standards are appropriate only when they can be tied to a cognitive component and can be referenced to a clearly stated criterion. For example, in the social information processing model noted above, the first step is to recognize the situation as one where a moral decision or political action is indicated. That means that the individual should process social cues in order to bring into a play a search for alternative solutions and behaviors. In one of the domains of interest in this chapter, the standard might deal with the individual's ability to process information about students of different races in their classroom in such a way so as to recognize their commonalities, or to process information about a problem in the community—for example, an environmentally polluting factory—in such a way that it would bring into awareness alternative points of view and a range of different solutions. Using a model such as this, the standard would not be set for the proportion of students who endorsed a particular attitude toward minority group students or the polluting factory, but whether the student could take the first step and process the information existing in the social environment in a way to understand different perspectives on the problem or to see the need for taking some kind of position together with like-minded members of the community. Other areas in which one might argue for such a cognitive anchor to an affective outcome might include understanding what it means for a society to operate under a rule of law as a cognitive step toward affective commitment to obey laws. Still another example would be understanding how cultures shape individuals' behaviors as a cognitive prerequisite to the affective characteristic of "international understanding". International research can be extremely useful in examining this position on the continuum, as it portrays a range of societies in which different patterns of social cues and ways of framing the problem may be found.

These alternatives (and others) should be debated in the context of understanding what it means to say that pupils have had the affective equivalent of an "opportunity to learn". One could even construct as process indicators the extent to which the climate of the school and classroom is such that everyone's opinion is treated with respect and open discussion is encouraged. Once again the developmental perspective is important. At what age is such classroom experience developmentally appropriate or necessary? International research has an especially vital role to play here as it pinpoints aspects of the context which favor the acquisition of important democratic values for which countries are interested in monitoring progress.

Laying out this continuum of alternatives relating to the non-methodological issues surrounding affective outcomes and standard setting raises more questions than it answers. But they are questions which, together with concerns for method, are important to consider.

References

Alsaker, F., & Flammer, A. (1994, July). Stress, future orientation, and well being of adolescents in 11 European countries and the U.S.: Methods and research. Paper presented at the symposium of the International Society for the Study of Behavioral Development, Amsterdam, the Netherlands.

Avery, P. (1992). Political tolerance: How adolescents deal with dissenting groups. In H. Haste & J. Torney-Purta (Eds.), *The development of political understanding* (pp. 39-51). San Francisco: Jossey Bass.

Board on International and Comparative Studies in Education (1993). *A collaborative agenda for improving international comparative studies in education.* Washington, DC: National Academy of Sciences Press.

Brislin, R. (1986). The wording and translation of research instruments. In W. Lonner & J. Berry (Eds.), *Field methods in cross-cultural research* (pp. 137-164). Beverly Hills: Sage.

Burstein, L., & Hawkins, J. (1992). An analysis of cognitive, noncognitive, and behavioral characteristics of students in Japan. In R. Leestma & H. Walberg (Eds.), *Japanese educational productivity* (pp. 171-220). Ann Arbor: Center for Japanese Studies, University of Michigan.

Cha, Y., Wong, S., & Meyer, J. (1988). Values education in the curriculum: Some comparative empirical data. In W. Cummings, S. Gopinathan & Y. Tomoda (Eds.), *The revival of values education in Asia and the West* (pp. 11-30). Oxford: Pergamon Press.

Center for Civic Education (1993, October). National standards for civics and government (Second and partial draft). Calabasas, CA: Center for Civic Education.

Dodge, K. (1986). A social information processing model of social competence in children. In M. Perlmutter (Ed.), *Cognitive perspectives on children's social and behavioral development* (pp. 77-126). Hillsdale, NJ: Lawrence Erlbaum Associates.

Fatygi, B., & Szymanczaka, M. (1992). *Report on Polish youth.* Warsaw: Wydawnictwo Interpress.

Gallhofer, I., Saris, W., & Schellenkens, M. (1988). People's recognition of decision arguments. *Acta Psychologica 68,* 313-27.

Gopinathan, S. (1988). Being and becoming: Education for values in Singapore. In W. Cummings, S. Gopinathan & Tomoda, Y. (Eds), *The revival of values education in Asia and the West* (pp. 131-146). Oxford: Pergamon Press.

Harber, C. (1991). Anti-racism and political education for democracy. In R.S. Sigel & M. Hoskin (Eds.), *Education for democratic citizenship: A challenge for multi-ethnic communities* (pp. 25-44). Hillsdale, NJ: Lawrence Erlbaum Associates.

Iudin, A.A. (et al.) (1993). Students' attitudes toward socialism, nationalism, and Perestroika. *Russian Education and Society 35,* 9-89.

Kuklinski, J., Riggle, E., Ottati, V., & Schwarz, N. (1991). The cognitive and affective bases of political tolerance judgments. *American Journal of Political Science 35,* 1-27.

Lee, G.B. (1990). Moral education in the Republic of China. *Moral Education Forum 15,* 2-14.

Nurmi, J. (1991). How do adolescents see their future? A review of the development of future orientation and planning. *Developmental Review 11,* 1-59.

OECD (1993). *Education at a glance: OECD indicators.* Paris: OECD, Center for Educational Research and Innovation.

Peschar, J. (1993, February). Education goals, the curriculum, and non-curriculum-bound objectives in OECD countries. Paper prepared for the Meeting of Network A, OECD/INES Project on Education Indicators, February, Vilamoura, Portugal.

Rest, J. (1983). Moral development. In J. Flavell & E. Markman (Eds.), *Handbook of child psychology.* (Volume 3, pp. 556-629). New York: John Wiley.

Saris, W. (1991). *Computer assisted interviewing.* Newbury Park, CA: Sage.

Schwarz, N., & Sudman, S. (1992). *Context effects in social and psychological research.* New York: Springer-Verlag.

Schwille, J. (1975). Predictors of between-student differences in civic education cognitive achievement. In J. Torney, A.N. Oppenheim & R. Farnen (Eds.), *Civic education in ten countries: An empirical study* (pp. 124-158). New York: Halsted Press of John Wiley.

Special Study Panel on Education Indicators (1991). *Education counts: An indicator system to monitor the nation's educational health.* Washington, DC: US Department of Education, National Center for Education Statistics.

Stevenson, H.W., & Stigler, J.W. (1992). *The learning gap. Why our schools are failing and what we can learn from Japanese and Chinese education.* New York: Summit Books.

Taylor, M. (1993). *Values education in Europe: A comparative overview of a survey of 26 countries in 1993.* Dundee, Scotland: Consortium of Institutions for Development and Research in Education in Europe.

Torney, J., Oppenheim, A.N., & Farnen, R. (1975). *Civic education in ten countries: An empirical study.* New York: Halsted Press of John Wiley.

Torney-Purta, J. (1992). Cognitive representations of the political system in adolescents: The continuum from pre-novice to expert. In H. Haste & J. Torney-Purta (Eds.), *The development of political understanding.* (pp. 11-25). San Francisco: Jossey Bass.

Torney-Purta, J., & Schwille, J. (1986). Civic values learned in school: Policy and practice in industrialized countries. *Comparative Education Review 30,* 30-49.

UNESCO, Principal Regional Office for Asia and the Pacific (1992). *Education for affective development: A guidebook of programs and practices.* Bangkok: UNESCO.

Voss, J., Tyler, S., & Yengo, I. (1983). Individual differences in the solving of social science problems. In R. Dillon & R. Schmeck (Eds.), *Individual differences in problem solving.* San Diego, CA: Academic Press.

Westin, P. (1989). *Assessment of pupil achievement: Motivation and school success.* Amsterdam: Swets & Zeitlinger.

Chapter 9

The Determination of Cut Scores for Standards

EUGENIO J. GONZALEZ and ALBERT E. BEATON

Boston College, School of Education, Boston, Chestnut Hill, USA

There is a growing interest among researchers and politicians in the monitoring of educational achievement at the local, national, and even the international level. This interest has given rise to research studies that have tried to measure educational performance across the world. These studies have reported their results in some type of scaling metric particular to the assessment instruments used. They also offered judgments on how well and how many of the students in a particular region meet certain education standards. These judgments on education standards are made by comparing the performance of the students on the assessment instrument with a cut-off score on the measurement scale. Several procedures for setting cut scores for standards have been suggested. These procedures and their advantages and disadvantages are introduced in this chapter. Other, more general issues related to the setting of cut scores for standards are also discussed.

It is not uncommon to hear people make statements such as "education standards need to be tightened up" or "we need to meet the standards". In these examples the term standard is understood to refer to a certain level of performance that is expected to be reached by a sector of the population of students. These standards are generally conceptual definitions that refer to levels of achievement, and they are usually set arbitrarily by a group of judges or other individuals that are considered to be either experts in the field or that have certain stakes in the outcomes being measured.

But how does one know that the standard has been met? The common way to find out if a standard has been attained is by measuring individual performance in the domain of interest. Such a measure is generally assumed to occur in a continuous dimension and on a unidimensional scale, and the standard to be reached is translated into a cut score on the measure. People who measure on one side of the cut score are thought to have reached the standard, and those measuring on the other side of the cut score are considered not to have reached the standard.

Several procedures have been recommended for selecting the cut score on an achievement scale, or for translating the standard into a score on the measure of the domain. This chapter will classify and describe the main procedures for the setting of cut scores. The adequacy of the use of cut scores and some general issues pertaining to their validity and usefulness are also discussed.

General Issues in Standard Setting

References to the monitoring of standards in education are widespread in the research literature and the mass media. Few people question the need to set standards in education. Such standards usually refer to a desired level of performance that is to be reached or—preferably—surpassed by most, if not all the students. They may also refer to a benchmark against which the achievement of an individual or group can be compared. This benchmark is then used to determine where on the achievement scale the individual or group would score.

The term standard is by no means unique to the field of education. There are standards in almost every other professional field. For example, there are known standards regarding the fuel efficiency of cars. These refer to the minimum number of miles per gallon that are deemed acceptable for a given car. This standard can be made more specific by indicating one standard for city driving and another for highway driving. So the standard of fuel efficiency may have two or more separate components that must be met by an automobile before it can be sold. The minimum required ratio of miles per gallon constitutes the cut off score for this standard of fuel efficiency. If a car's does not meet this standard—i.e. is not fuel efficient according to the standard—then the model may be taken off the market or be redesigned at the factory.

The above example illustrates one way of conceptualizing and using a cut score. This is what is sometimes called the "all or nothing" approach. A cut score is set so that below this point the individual or object measured is disqualified in some way and above it the individual or object is qualified. This is the case when the cut off score is set for a mastery situation. But there is another approach to setting cut scores. Consider that the fuel efficiency of cars is to be compared in two different states. One way to do this would be to compute the average fuel efficiency of the cars in each of the states and then compare the two mean scores. But, by using only the mean score important information about the characteristics of the distribution is lost. Extreme values in the distribution have a significant effect on the mean score and hence may disguise important differences or exagerate trivial ones. Alternatively, instead of computing the mean fuel efficiency one could, according to pre-specified criteria, classify the cars in each state according to four levels of fuel efficiency: low, medium, high, and excellent. For each of these levels a cut score in the measure on fuel efficiency would be set, so that the fuel efficiency of each of the cars could be compared and classified in the corresponding category.

This example of fuel efficiency illustrates several points about the setting of cut scores in education. Before a cut score can be defined, a measure on which to compare the "fuel efficiency" of each car must be found. This measure needs to be reliable and, even though it may seem as simple as filling up the tank, driving the car around, taking note of the miles driven, and then see how much of the fuel tank was consumed, there are several issues that need to be carefully considered. The kind of driving that was done with the car is a relevant variable; the temperature at which the car is tested and frequent acceleration will make a difference in how much fuel is consumed. A good measure of fuel efficiency is also needed. Should the measure be what goes into the gas tank or what actually goes into the engine? In other words, a reliable and valid measure is required to determine the performance of the car with respect to fuel efficiency.

Once the measure is established, the second important issue then becomes where the cut score should be set. How much of the measure of fuel efficiency is enough to label the car as being fuel efficient? One might expect that a group of experts would be called upon to determine how much fuel efficiency is enough. Several factors may be considered in setting this standard. These could be environmental issues, availability and cost of fuel, distance to be traveled by the car, etc. Many of these criteria may be measurable in detail, whereas some others may just be subjective judgments from experts in the field or other people who have an interest in these issues and their consequences. But the final decision on a cut-off point depends on the guiding principle behind the decision and the judgment of the people making it. A concern for preserving energy and lowering pollution levels may lead to a "tough" standard. A concern to keep profits up in the oil industry may keep the standard at a "low" level. Ultimately, a decision is reached based on a compromise between the two or more competing or collaborating interests. In some cases the decision is made at the expense of one of the principles. In complex cases a mathematical model may be used to arrive at a weighted standard. The decision as to where a cut score should be set then depends on the choice of the factors to be entered into the model and on the relative weight given to each factor.

Purposes and Uses of Cut Scores

A cut score can be considered as a point on a scale that is used to sort examinees into two or more categories that reflect different levels of proficiency relative to the particular objective measured by the test items on which the scale is based (Hambleton, 1978). Methods for selecting cut scores may be based on considerations of item contents, educational consequences, psychological and financial costs, performance of others, and errors due to guessing and item sampling. Each of these methods may yield different cut scores. The selection of a method depends on the circumstances and the purpose for which the cut scores are to be used, how much time is available to set them, what resources are available, and how capable the individuals who need to apply the particular method are. The task of setting a cut score thus may sometimes be preceeded by extensive in-service training.

When selecting cut scores for education standards it is important to consider their purpose and use. Cut scores can be set for different purposes, such as individual diagnosis, certification, and program evaluation. They can become helpful indicators of whether a student has learned a topic and is ready to go on to the next. Unlike when used with individual tests, in program evaluation the data can be aggregated and an appropriate benchmark can be used to attach value to the results. But pass/fail cut scores may not always be the best yardstick for judging what happens to groups at different locations on the performance continuum or for diagnosing program strengths or weaknesses. In these cases more than one cut score may be necessary.

Standards were traditionally set for the purpose of certification. Since the mid-1980s, however, some governments have become increasingly interested in the setting of standards that allow to them determine how well their education system is doing. For example, in the United States the president and the governors met in a historical meeting that took place in Charlottesville, Virginia and came out with a document entitled "America 2000: An education strategy". Among the national education goals listed in this document, the third one indicates that "American students will leave grades four, eight, and twelve having demonstrated competency in challenging subjects ..." (America 2000, p. 3). Even though loosely defined, this is a standard that is to be met by all students in the country. But if mathematics is taken as a challenging subject, and if the students are to be competent in this subject at all three grades, then the implied standard must refer to different levels of competence, since it is probable that competence is defined differently depending on the grade considered.

The issue of setting standards and selecting cut scores has been present in the field of education ever since evaluation has been systematically conducted (see Chapter 1). But the issue did not become a subject of research until the mid-1960s. It is not difficult to imagine our early ancestors teaching their children to fish and hunt, and somehow setting a standard that had to be met before the children were allowed to go out on hunting expeditions. Or in agriculture, the time when a crop has to be harvested is often based on a standard of coloring and texture that the fruit or vegetable has to meet.

In the 1960s, with the growth of criterion referenced testing and the proliferation of testing for certification in the United States, the issue of selecting a cut score was given special attention in the research literature. This interest was sparked by the civil rights movement that emerged at that time (Hills, 1971). A more equitable selection system was demanded by various groups in the North American society and, hence, there was a need to make the process of selecting cut scores a standardized and rational one. This led to an effort to develop procedures for the setting of cut scores that would be justified in light of the education theories and political goals of the period.

Cut scores were at first used for the purpose of selection—for graduation, certification, licensing examinations, and even for jobs. Standards were treated as a dichotomous variable where the individual either reached the standard or failed to reach it, and a selection decision was made based on the individual's performance. In contrast, the emerging trend has been for the selection of cut scores for standards to occur in a somewhat different context. Standard setting is now advanced as a tool of monitoring the educational progress of groups of individuals, where the consequences of meeting the standard or not are decided not at the individual but at the group level.

The standards and cut scores set by the National Assessment Governing Board (NAGB) of the US National Assessment of Educational Progress (NAEP) can be.considered as an example of this change. In 1992, the NAEP began reporting cut scores on its assessment scales that were to categorize the performance of the individuals taking a test into four categories of achievement: below basic, basic, proficient, and advanced. For this purpose, three cut scores are set along the achievement scales for the 4th, 8th and 12th graders that are tested. The NAEP is not designed to report at the individual level, and so individual students are not classified. Instead, the classification of results by the levels of achievement indicated by the cut scores is intended for use by policymakers and researchers for the purpose of making group decisions.

Cut scores for standards have also been used to compare the performance of different groups, as in the NAEP Trial State Assessment, where students in over 40 US states were tested in mathematics. For each state the results were summarized by the percentage of its students meeting each of the proficiency levels mentioned above. Even though some people argue that essential information is lost if the results are categorized into four discrete categories, there are also some advantages. For example, one may be interested in finding out if the students in a certain population meet a given standard. It then becomes irrelevant how far beyond the standard the students are. Such is the case when qualifying times are set to determine who is to participate in an athletic competition. The individual or team must perform at a certain level in order to be admitted into the competition. A level of performance well above the cut score allows the athlete to participate no more than a performance just above the cut score.

In summary, the setting of cut scores for education standards generally serves two purposes: selection and description, which can occur at the individual or group level. The difference between these two approaches is not trivial, since it has important implications in terms of how the cut score is to be selected and what the implications will be for the individuals and groups to be classified with it. For example, a cut score for selection is that which is used to select job applicants or to admit prospective students into an education program—whether this be a remedial program or an advanced honors class. A cut score for description is one that will be used to describe a population in terms of meaningful categories or to determine the effectiveness of an education program. This is the case with the cut scores set along the NAEP scales mentioned previously. These points were selected so as to summarize the performance of the students on the achievement tests. But instead of reporting the results in scaled scores, the selection of the cut scores makes it possible to describe the student population in terms of the percentages who have reached meaningful education standards. In this case, the setting of cut scores can be considered less consequential for the individual student than when standards are set for the purpose of selection.

Another difference between a cut score for selection versus one for description is that of the importance given to the error of measurement when setting the score. If a cut score is used to describe a population, then one can be less uncomfortable about erring either way in the classification, assuming that the errors will cancel out at both sides of the cut score. But if a cut score is used to make decisions about individuals, then making an error on either side may have severe, and unwanted consequences. Thus the important issue of classification error must be considered, as must the assumptions made when setting a cut score.

Criticism of Standard Setting Procedures

The procedures used in the setting of standards and the selection of appropriate cut scores have been often criticized. The argument that standards are set following arbitrary procedures has generated much concern and controversy. This issue of arbitrariness therefore needs to be addressed in this chapter before the procedures are described. In the Webster's Ninth New Collegiate Dictionary, the term "arbitrary" is presented as having three generally understood meanings. The first one refers to "something depending on individual discretion (as of a judge) and not fixed by law". The second meaning of the word arbitrary is "not restrained or limited in the exercise of power". The third meaning—the one that is often and unfortunately attached to the selection of cut scores for standards—understands the term as "something based on or determined by individual preference or convenience rather than by necessity or the intrinsic nature of something; existing or coming about seemingly at random or by chance or as a capricious and unreasonable act of will".

Popham (1978) indicates that while it can be said that performance standards must be set with the use of judgment, it is patently incorrect to equate human judgment with arbitrariness in its negative sense. Linn (1978) adds that the problem of standard setting is not the same as that of creating minimum competency or criterion-referenced tests. There were standards for graduation or for passing from one grade to another long before these terms appeared in scholarly journals and even long before such journals existed. Even without state-wide, district-wide or school-wide testing, teachers have always had to decide on standards for passing and for assigning grades in one form or another. However, the problem of how cut scores should be set, as well as issues of equity and fairness, has brought the matter into the spotlight.

An example of the criticism advanced against the setting of standards and the selection of cut scores on achievement tests is offered by Glass (1978):

> I have read the writing of those who claim the ability to make the determination of mastery or competence in statistical or psychological ways. They can't. At least, they cannot determine 'criterion levels' or standards other than arbitrarily. The consequences of the arbitrary decisions are so varied that it is necessary either to reduce the arbitrariness, and hence the unpredictability, or to abandon the search for criterion levels altogether in favor of ways of using test data that are less arbitrary and hence safer (Glass, 1978, p. 237).

Glass' use of the word arbitrary seems pejorative. The selection of a cut score is always arbitrary to some degree. What should be sought is an informed procedure with numerous pieces of evidence that will support the selection of a given cut score. This means, among other things, looking at the consequences for different groups of test takers when the cut score is selected, verifying the adequacy of the scale on which the cut score is placed, and studying the validity for different purposes. Careful consideration of the consequences that the selection of a cut score may have for different groups in a population is particularly important in the validation process.

In general, the definition of a cut score for a standard starts with the selection of a group of people who determine what would be expected of the person or object being assessed, how that assessment would occur, how much error will be tolerated, the number of criteria to be used in making the decision, and last, but not least important, where the cut score for each of the criteria will be placed. These steps can be performed by the same or by different groups of people.

There are different procedures currently available for setting cut scores for standards. When applied to the same standard and scale of measurement, they do not always coincide in their results. Given that different procedures are used under each of the methods, and that they have different assumptions, it is not surprising that they yield different results. Instead of this being a deterrent, it should focus attention on the choice of the appropriate method for the task at hand. More than one method may be used and their results combined, and various pieces of evidence can be examined in order to confirm the adequacy of the applied methods.

The lack of clear, unequivocal rules for the setting of cut scores is, for many, a sign that the whole issue should not be pursued further. For others, it points to the need for further exploration and research, as well as careful use of them.

Assumptions in Selecting Cut Scores

Because the procedures used in selecting cut scores for standards are imperfect, the assumptions and circumstances under which the cut scores were obtained must be taken into account in interpreting the results. Six crucial assumptions often made when selecting a cut score are discussed below.

Measurability. The first assumption is that the construct of interest—or its manifest variables—can be measured either directly or indirectly. In education the measure is generally a test, which consists of one or more test items relevant to a certain domain. It is generally assumed that a test can be constructed on the basis of a set of questions, and that the generic score derived from the answers to these questions indicates how well the individual can perform in the specified domain. This assumption limits the development of standards to measurable domains or to measurable aspects of those domains.

Validity. The second assumption is that what is measured is relevant to the desired outcome or standard. This assumption has to do with the validity of the measure (see Chapter 8). It is generally assumed that the test—and the associated scale on which the cut scores are set—is relevant to the construct one seeks to measure. For example, when using a test to select applicants for university admission, it is assumed that the performance of the prospective students on the test is directly related to the level of performance these students would have if they were enrolled at the university.

Validity, as used here, is defined in accordance with Messick (1989, p. 13): "Validity is an integrated evaluative judgment of the degree to which empirical evidence and theoretical rationales support the adequacy and appropriateness of inferences and actions based on test scores or other modes of assessment".

Reliability. A third assumption is that the domain is measured with a known and acceptable amount of error. This assumption relates to the reliability of the measure of the domain. A test score gives an indication of the performance on the domain with an estimated variability or error. The recognition of the amount of error in the measure is crucial, particularly if a cut score is to be used to make decisions on individuals. An error in either direction of the scale may have severe consequences. In contrast, if decisions are made about groups then it can be assumed that the errors will cancel each other out. There are two types of errors that can be made when a cut score is selected. A false negative is when an individual is classified as not having met the standard according to the measure, but in reality the individual does possess the trait in the required amount. A false positive is when an individual surpasses the cut score for the standard but the performance in the domain is inferior to the one indicated by the performance on the measure.

Consistency. A fourth, and sometimes ignored assumption is that the relationship between the measure and the domain is the same—or different within reasonable limits—for all individuals that are compared against the standard. In other words, it is assumed that the standard classifies with equal precision and validity individuals from each of the different groups being tested. That is, individuals with the same capability will be consistently placed on the same point along the measurement scale, regardless of the differences in personal and contextual characteristics that are unrelated to the domain being measured.

Unidimensionality. The fifth assumption concerns the supposition that the individuals are compared along a unidimensional scale that corresponds directly with their performance in the domain being tested. The setting of a cut score implies that the individuals are placed along a unidimensional scale and that the cut score represents a break in this continuum above which certain behaviors are expected and below which they are not. This assumption is behind the contribution of item response theory (IRT) to the selection of reasonable cut scores. In theory, when the assumptions of IRT are satisfied, it allows for the construction of unidimensional scales. In practice, however, the assumption of unidimensionality in IRT models has been questioned. It should be added that under IRT unidimensionality is defined statistically rather than psychologically (Bejar, 1983). Hence unidimensionality—as defined under IRT—does not imply that the performance on the test items used to set the cut score is due to a single psychological process; in fact, a variety of psychological processes are acknowledged to affect the response to a given set of items. Unidimensionality, in this case, refers to the fact that the responses to the test items can be described with the use of a mathematical model that assumes them to be generated or affected by one single dimension, regardless of the psychological or cognitive dimensions actually involved in the construct.

Interpretability. The sixth and last assumption is that the test score is interpretable as a sign of trait dispositions or internal states, as samples of domains or response classes, or as some combination of the above (Messick, 1989). This is, it is assumed that the score is directly related to the domain of interest.

Interpretation of Cut Scores

There is a difference between a standard and its cut score. The standard is the theoretical definition of what is expected from a student. For example, the performance level required to be competent on a given job. The cut score is the operationalization of this standard onto a numeric scale. Inferences are made from the scale and the cut score to the standard. It is assumed that reaching the cut score or surpassing it means that the standard has been reached or surpassed. This is important when interpreting cut scores set for standards, and it allows for the use of the error of measurement in interpretations. If a score is seen as an indicator of a trait or internal disposition, then the cut off level compared to an individual's score is an indicator of whether the individual has met the standard. A level of performance above the cut score is considered as a sign that the individual is likely to have met the standard, and a performance below the cut score shows that the person is likely not to have met the standard.

Cut scores can also be interpreted in different ways, depending on the model used in developing the measure on which it is located. The two models considered here are classical test theory and IRT. Under the assumption of classical test theory the test items may be considered a random sample from a theoretical domain. The score on the test as a whole, or the proportion correct, is then considered as an estimate of how an individual would have done if all of the items in the domain had been administered. The standard error of this measure refers to the variation on the score obtained by the individual if numerous random samples of the same number of test items from the domain had been administered. The test score is assumed to be the best available estimate of the true score for that individual.

Under the assumptions of IRT, the interpretation of a score is different. The items are assumed to be located along a difficulty scale, and the score of the person on the set of items administered is considered to be the best indicator of where the person is located along the scale. This location is called the ability of the person. This location is assumed to be the same regardless of the difficulty of the items responded to by the examinee. The error of measurement is the variation in the estimated ability for the individual if numerous sets of items were administered to the same individual and ability estimates were to be obtained with each set of items. In classical test theory, the percentage of correct answers by an individual is directly related to the difficulty of the items. Under IRT assumptions, the ability of the individual is independent of the difficulty of the items. A person with the same ability will have different percent correct scores depending on the difficulty of the set of items administered, whereas this same person would have the same ability level. These differences between classical test theory and item response models favors the use of IRT when designing tests that will be used for setting a cut score, since a scale that is independent of the sample of items and the particular sample of people taking the tests is desirable.

The notion of the error of measurement in the selection of a cut score deserves special attention. This is so because the decision based on a cut score can be no more precise than the measure on which the cut score has been placed. Sometimes a certain amount of error of measurement is used to adjust the cut score in a given direction, particularly when the score is used to make decisions about individuals. For example, Ravelo and Nitko (1979) used two cut scores on an admissions test to select students at a university. One cut score was located at a place where it was considered very certain that the student would perform adequately in the institution. The second score was placed where it was fairly certain that the student would not perform adequately. All the applicants scoring above the highest cut score were accepted and those scoring below the second one were rejected. The people in between these two cut scores were accepted to the institution using a lottery system with a probability of selection based on their score. Students from this group were selected until all places at the university were filled.

In other cases, the cut score is selected by adding or subtracting a certain amount of the standard error of measurement to the originally selected cut score. However, this practice of adjusting a cut score based on the standard error of the measure can be questioned. There is no agreement on issues such as how much the score should be adjusted, and in which direction. Furthermore, the fact that a score is adjusted downward does not make it any more reliable, although more people would be considered to have reached the standard. A common way of adjusting a cut off score is by adjusting it downward by either one or two standard errors of measurement. But, again, there are no agreed guidelines for whether or how much to adjust the score.

One last concept to be reviewed before describing the methods for setting cut scores is that of criterion referenced tests. A criterion referenced test is defined as one that consists of a sample of tasks from a well-defined population of tasks. This sample of tasks is generally in the form of test items and is used to estimate the level of performance in that population at which the individual can succeed (Hambleton, & Eignor, 1989). As indicated by Jackson (1970, p. 3, cf. Glass, 1978, p. 242), the "term criterion referenced [is] used here to apply only to a test designed and constructed in a manner that defines explicit rules linking patterns of test performance to behavioral referents".

Most of the procedures described below, except for those based exclusively on statistical analysis, start with the selection of judges who decide what the cut score will be, given a defined standard. Not an exact science by any means, but one that can be standardized. This points to the importance of the training given to judges, as well as the judges' background and expertise.

Classifications of Standard Setting Procedures

The procedures used to set cut scores for standards can be classified in different ways. Such classifications can illuminate some pertinent issues in the selection of a particular method.

All procedures begin with defining the purpose of the standard. As mentioned previously, a standard is normally determined in order to describe or classify a population. A clear definition of the domain to which the inferences are made is also needed. That is, the kind of performance to be considered needs to be carefully determined. The next step concerns the translation of the domain-specific standard to a cut score. The various ways in which this can be done are classified and explained below.

Glass' Procedural Classification

A useful classification of standard setting procedures has been proposed by Glass (1978). He identified six categories of procedures for determining the criterion score on a criterion referenced test: (1) comparison-group approach; (2) mastery approach; (3) bootstrapping approach; (4) minimal-competency approach; (5) decision-theory approach; and (6) operations-research approach.

Comparison-group approach. Comparison-group procedures are based on comparisons of the performance of one group with that of others. This group comprises all procedures that compare the performance of unclassified people on the criterion measure to the performance of other people on the same criterion measure, where the latter have been previously classified according to a different criterion. Consider the classical case when the performance of apprentices on a test is compared with that of professionals on the same test. Several procedures can be used to arrive at a cut score. A common one employs a measure of central tendency from the professional group to set a cut score for selecting apprentices. Such procedures assume an empirical approach, since the score is set on the basis of the level of performance of others, rather than on a theoretical decision by a group of judges.

Procedures of this sort have several serious drawbacks. The motivation to perform well on the test may differ considerably between the two groups. Moreover, the procedure does not account for those professionals already in the field who perform below the measure of central tendency that is selected as the cut score. It is not clear, then, how these professionals would score if they had to retake the test. In some cases people's professional records do not match the results on the criterion measure. Another issue relates to the reliability of the criterion, and what one should do with borderline scores. In conclusion, the performance of others is geneally considered inappropriate when used as the sole criterion for selecting a cut score because the comparison group is mostly imperfect.

Yet, despite the drawbacks, comparison methods can be useful under certain circumstances, such as when there is no need to set a theoretical standard because an empirical standard is deemed sufficient. Those procedures that set a cut score for a standard based on a quota can also be considered as belonging to this class of methods. Quota methods are commonly used in selecting applicants to an education program, for example a course in business administration. Knowledge about the level of performance attained by others may be valuable information in this case. For example, the standard cut score can be set as the median score of those who previously passed the same course with an above-average score. Another case is when a quota is used to set the standard. For example, the top-scoring 1500 students may be accepted rather than using some other, theoretically based approach.

Mastery approach. The mastery approach relies on counts backward from a "100 percent correct" performance level. When a mastery test is constructed it is sometimes assumed that all of the included items are relevant to the domain and, consequently, that all items should be answered correctly by any person who meets the standard. But in order to account for error of measurement and other errors that influence the score of a person, the cut score is set back from 100 percent correct to a lower, but still acceptable level. This may apply, for example, in the case of a drivers examination, when 20 or 30 items are administered and the examinee is allowed a few incorrect responses. Of course, it would still be expected that any driver should know the traffic rules, but there may be factors that affect the respondent and so the minimum acceptable score is adjusted to account for these unwanted influences.

The use of this approach assumes that there are two possible states: mastery and non-mastery. Hence it cannot be employed if more than one cut score must be selected, as would be the case when students are to be selected for an advanced, an intermediate or a beginners class, unless the individual takes a test specific for each level of performance. The procedures under this approach rely heavily on classical test theory, since "percent correct" is the score resulting from a test. Accordingly, the performance of an individual on the measure of the standard depends on the difficulty of the test items that are sampled from the domain.

Bootstrapping approach. Under this category fall the procedures that compare the performance of the students on a measure under development with an external criterion. This is the case when a mastery examination is being developed and the selected criterion is based on the performance of the individuals on another test.

A drawback is that the reliability of each of the measures is generally unknown. Unless the tests are parallel, the results on the two tests do not correspond precisely. Glass (1978) doubts whether this approach is reasonable: if the cut score on one test is selected on the basis of the performance on a second test, then why not employ the latter and be done with it? However, this last option does not always apply. For example, when a new test is developed to substitute a previous one, or when cut scores are to be chosen on parallel tests, then the performance on each of the tests may need to be made comparable by selecting equivalent cut scores.

Minimal-competency approach. Procedures involving judgments of minimal competency are used frequently. They are sometimes also referred to as procedures that rely on the judgment of items rather than of people. Angoff's method (Angoff, 1971); Nedelsky's method (Nedelski, 1954), and Ebel's method (Ebel, 1972) are among the principal ones in this category. These and other methods will be introduced below; they are discussed in more detail in the next chapter.

Several issues pertain to methods that require judges to make decisions on what the performance of a minimally competent person should be like. Then, based on the decision, and sometimes in combination with a mathematical procedure, the judged performance on the test is translated into a cut score. The use of judges to decide how a minimally competent person would perform on a test points to the importance of selecting judges that can make such decisions reliably and fairly.

Glass (1978) doubts whether judges can actually make reliable decisions. He also questions whether the judges can be expected to define the notion of "minimal competency". Nonetheless, expert judgments are widely used in large scale assessments of student performance.

One advantage of this approach is that the selected standard is not based on the performance of others but on a theoretically defined performance level. Another advantage is that the approach allows for several levels of performance to be selected on the same measure. That is, several standards can be linked to a scale and their corresponding cut scores selected.

Decision-theory approach. Several procedures rely on applications of decision-making theory. Under these approaches the persons are categorized into two groups according to certain external criteria, and the proportions in each group are computed. The cut score on the external criterion is assumed to be fixed whereas the cut score in the test is varied until the wrong classifications are minimized on the external criteria. This principle undergirds the use of a discriminant analysis function for determining an optimal cut score. The dependent variable is the external criterion and the independent variable is the measure on which the cut score is to be placed. It can be assumed that the cost of wrong classifications is either the same for false negatives and false positives or favors one type of wrong classification over another, less desired one. These procedures are useful if reliable external criteria are available, but this is unfortunately seldom the case.

Operations-research approach. This is the last group of standard-setting procedures Glass (1978) discussed. The general principle is that of maximizing a valued commodity by finding an optimum point on a mathematical curve or graph. A mathematical function describing the relationship between two or more variables of interest and the cut score needs to be defined. The score chosen is that which satisfies certain pre-established criteria. Again, there is a reliance on the reliability of the measures used to set the standard, and the corresponding cut score is set empirically rather than on a theoretical basis. The approach assumes that there are external measures that can be used to carry out the procedure adequately.

Burton's Procedural Classification

Burton (1978) offers a different classification of procedures to set cut scores. He distinguishes between procedures for standards that are based on theories, standards based on expert consensus, and standards based on practical necessities.

The first group refers to procedures that assume the use of a learning hierarchy theory. The final score of a respondent on a test measuring performance on a learning hierarchy can be used to infer what that person can do or cannot do along the entire hierarchy of tasks. The score implies that the respondent can do the corresponding task and all other tasks that rank below it on the hierarchy. The adequacy of this approach depends on how well the theory or model can be measured. Hence it should be used only if the hierarchical nature of the performance domain has been validated empirically.

Standards can also be set based on the experience of a panel of judges. Such standards involve comparisons between the test items and reality. The justification and the consequences of the standard must be established externally, since they do not necessarily follow logically from a theory. Burton (1978) notes: "... the irony with the techniques of using experts to set standards is that the whole purpose of proposing criterion-referenced testing is to take educational decisions out of the hands of individuals and place them in the machinery of objective, scientific techniques".

The last group of procedures described by Burton are those for standards based on practical necessities, such as quotas or prerequisites for personal efficacy. This is generally the case with procedures that use the cut score in yes and no situations.

Like Glass (1978), Burton rejects all three groups of procedures on philosophical grounds. Yet he also admits that cut scores can be useful in helping educators to make decisions.

The Priori and Posteriori Distinction

A third classification of methods for setting cut scores for standards distinguishes between two groups: *a priori* and *a posteriori* methods. *A priori* methods select the cut score prior to the test administration. In contrast, *a posteriori* methods apply only after the test data have been collected. This classification in many ways parallels the one based on a judgment of items (*a priori*) and of people (*a posteriori*).

A priori Methods

A priori methods are generally based on judgments about the difficulty of the test items for a certain group of individuals. Several methods fall into this category. Among these are Angoff's method (Angoff, 1971), modified Angoff's method (Angoff, 1971), Ebel's method (Ebel, 1972), Jaeger's procedure (Jaeger, 1989), and Nedelsky's (Nedelsky, 1984) procedure. These methods, which are introduced briefly below, are described in more detail in a subsequent chapter.

Angoff's method is best described in his own words: "A systematic procedure for deciding the minimum raw scores for passing and honors might be developed as follows: keeping the hypothetical acceptable person in mind, one could go through the test item by item and decide whether such a person could answer correctly each item under consideration. If a score of one is given for each item answered correctly by the hypothetical person and a zero score is given for each item answered incorrectly by that person, the sum of the item scores will equal the raw score earned by the 'minimally acceptable' person" (Angoff, 1971, pp. 514-515).

The crucial difference between Angoff's method and the one known as *modified Angoff* is that in the latter the judges are given a list of probability values for each item before they are asked to choose the one that most closely approximates their own, subjective estimates. *Ebel's procedure* (Ebel, 1972) follows quite similar steps. It requires the judges to assign a probability value to the likelihood that a minimally competent person would answer a given test item correctly. All test items are classified into a matrix comprising 12 cells and two principal dimensions. The first relates to the difficulty of the items and the second to their relevance. The judges must assign each test item to one of the 12 cells in the matrix. They must then guess the proportion of the items that would be answered correctly by a minimally competent person—assuming that this person would take a large number of similar test items.

Jaeger's procedure can be described as follows: "Each judge in a panel must answer the following yes-no question for each item on the competency test: Should every examinee in the population of those who receive favorable action on the decision to be based on the test be able to answer the test item correctly?" (Jaeger, 1989; p. 494). The procedure "is iterative, in that judges are given several opportunities to reconsider their initial judgments on the items. Prior to reconsidering their recommendations, the judges are also given data on the actual test performance of the examinees under consideration, in addition to information on the recommendations provided by fellow judges".

Nedelsky's procedure (Nedelsky, 1954) can only be used with multiple choice items. A main difference between this procedure and the previous one is that instead of setting the score of the 'minimally competent' examinee, it selects the score that would most likely be obtained by that person if relevant distractor variables were held constant. The judges are asked to predict the behavior of the minimally competent examinee on each of the options of a multiple-choice test item. In particular, the judge is asked to specify which of the distractors in the item a minimally competent examinee would be able to recognize as incorrect and eliminate as a possible correct answer. A statistic called the minimum pass level is then computed for each item. The minimum pass level is the reciprocal of the number of item response options that the examinee would not be able to recognize as incorrect. The mean value of the sum of the minimum pass levels on each of the test items and for each of the judges is then used as the cut score.

A Posteriori Methods

These methods can only be implemented after the test data have been collected. A cut score is set based on the actual rather than hypothesized test performance of a group of examinees. The borderline group procedure and the contrasting group procedure are among the better known ones in this category.

The *borderline group procedure* is described by Livingston and Zieky (1977). It involves the setting of cut scores based on the qualifications of those taking the test. Judges are asked to define the level of knowledge or skill that a person must possess to be regarded as competent. They are specifically asked to define three categories of competence: adequate, borderline or marginal, and inadequate. The test is administered only after all the respondents have been classified in one of these categories on the basis of information other than that person's test score. The resulting cut score is the median of the distribution of competency scores achieved by the examinees who are classified as borderline.

The *contrasting group procedure* was also proposed by Zieky and Livingston (1977). It is conceptually similar to a standard setting method proposed previously by Berk (1976). Again, the focus is on the perceived competence of the examinees rather than on the difficulty of the test items. Judges are identified and a sample is drawn from the test population. Categories of competence are defined and individuals are assigned to these categories. A cut score is then determined according to an analysis of those who were classified as either competent or not competent.

Other Issues in Setting Cut Scores

In education, performance tests are generally used to estimate examinees' domain scores and to assign them to mastery states. However, the persons located near and on both sides of a cut score will often be indistinguishable, since there is always error attached to a cut score. Standard setting methods therefore normally require that additional information be gathered and analyzed, so that the arbitrariness of the decision can be reduced. The more reliable the cut score, the more valid the distinction will be between those who are located well below or well above the standard. Koffler (1980) therefore recommends the use of a mix of standard setting methods, since this can bypass some of the limitations associated with the use of only one method.

The amount and type of information given to those who rate the test items must also be considered. The judges can be given empirical information on the relative or the absolute difficulty of the items. In the first case, the judges are given a set of items ranked by their difficulty level. In the second case either the IRT-estimated difficulty level or the "percent correct" on each test item is provided. In a seminal study, Lorge and Diamond (1954) rated the performance of judges who made decisions on the difficulty of test items. They classified the judges into three groups according to the accuracy with which they were able to determine the difficulty of the items. The judges were then given information about the actual difficulty of the items. The authors concluded that providing the judges with information about the absolute difficulty of the items improved the accuracy of subsequent ratings and hence the reliability of the resulting cut scores. This is the reason why this approach is taken in large-scale assessments such as the NAEP in the United States.

A further issue arises if more than one measure is used. In this case one may be led to set a cut score in each of the measures, so that those who reach or surpass the cut scores in both measures are classified as meeting the standard. But, as Lord (1962) points out, the use of multiple cut scores may be inappropriate if the selection measures are fallible. In his study, Lord investigated the shape of an optimum selection region. He concluded that—other things being equal—the lower the reliability of the predictor variables, or the higher the correlations among them, the greater the discrepancy tends to be between the multiple cut score selection region and the optimum selection region.

Most of the methods described in this chapter were originally developed for the purpose of selecting a single cut score on a single measure. But the proliferation of assessments for the purpose of accountability and program evaluation has led to a demand for multiple cut scores on a measure. For example, one may be interested in using cut scores to separate a population into groups with basic, intermediate, and advanced skills. Two cut scores will be required in this case. Such cut scores can be obtained using standard procedures if specific instructions are given to the judges. Instead of requesting them to imagine a minimally competent person, they will be asked to imagine, for example, a person with only basic skill in a given subject.

The use of new IRT models in the test development and scoring process makes it possible to derive empirically information relevant to the decision whether a standard has been met. IRT models offer three useful indicators. First, the *item information function* can be used to increase the precision of measurements around the cut score, because the inverse of this function gives an estimate of the error of measurement of an item along the entire ability scale. Hence it is possible to select the items that have a small error variance around the point where the cut score is to be set (Hambleton, 1989). If more than one cut score is needed, then IRT models make it possible to select those items that have a small error variance at different points along the scale. The *item discrimination function* in an IRT model provides another useful indicator in the selection of test items. It estimates the probability of a correct response to an item as a function of a change in the ability of the individuals responding to that item. Items with high discrimination are preferable to items with low discrimination. The third indicator is the *guessing index*. This is a parameter of the probability that an individual with low ability is able to answer an item correctly. Items with low values on the guessing index are preferred over items with high guessing parameters.

Conclusion

This chapter has dealt with many issues in the setting of cut scores for educational standards. Although a mainly technical approach has been taken in this chapter, it should be noted that the decision to set a cut score is principally a political matter. This decision is always an arbitrary one because it is ultimately based on value judgments—albeit judgments that are informed by technical considerations. The decision to set a cut score for a standard is not fixed within law but depends on the individuals who select the methods, develop the items, design the scale, and who eventually determine the cut score itself. The currently available methods for the setting of cut scores for standards are therefore only tools that assist in answering the questions posed by individuals. Hence these questions must always be carefully posed and examined.

According to the research literature to date, the most widely used method seems to be the one proposed by Angoff (1971). However, many other standard setting methods are also sufficiently flexible and can be adapted to different circumstances. But, despite this, all methods have at least some disadvantages. The purposes of their application and the outcomes of their use should therefore be carefully considered at all times. In particular, the consequences to the individuals being classified should be taken into account when selecting a method and setting a cut score.

A vast number of studies on the setting of cut scores exist in the research literature. An issue of the *Journal of Educational Measurement* (Volume 15; 1978) is of interest because it contains a number of excellent articles on the subject. Not less important are the writings by Jaeger (1976), Meskauskas (1976), Zieky and Livingston (1977), Shepard (1980), and Berk (1986). Lastly, the reader is encouraged to turn to the next chapter, which takes the topic an additional step further.

References

Angoff, W. (1971). Scales, norms, and equivalent scores. In R.L. Thorndike (Ed.), *Educational measurement. Second edition.* New York, NY: American Council on Education and Macmillan Publishing Company.

Bejar, I.I. (1983). Introduction to item response theory and their assumptions. In R.K. Hambleton (Ed.), *Applications of item response theory* (pp. 1-23). Vancouver, Canada: Educational Research Institute of British Columbia.

Berk, R.A. (1986). A consumer's guide to setting performance standards on criterion referenced tests. *Review of Educational Research* 56(1), 137-172.

Block, J.H. (1978). Standards and criteria: A response. *Journal of Educational Measurement* 15(4), 291-295.

Burton, N.W. (1978). Societal standards. *Journal of Educational Measurement* 15(4), 263-272.

Ebel, R.L. (1972). *Essentials of educational measurement.* Englewood Cliffs, NJ: Prentice-Hall.

Glass, G.V. (1978). Standards and criteria. *Journal of Educational Measurement* 15(4), 237-262.

Hambleton, R.K. (1978). On the use of cut-off scores with criterion referenced tests in instructional setting. *Journal of Educational Measurement* 15(4), 277-290.

Hambleton, R.K. (1989) Principles and selected applications of item response theory. In R. Linn (Ed.), *Educational measurement. Third edition.* New York, NY: American Council on Education and Macmillan Publishing Company.

Hambleton, R.K., & Eignor, D.R. (1978). Guidelines for evaluating criterion-referenced tests and test manuals. *Journal of Educational Measurement* 15(4), 321-327.

Hills, J.R. (1971). Use of measurement in selection and placement. In R.L. Thorndike (Ed.), *Educational measurement. Second edition.* New York, NY: American Council on Education and Macmillan Publishing Company.

Jaeger, R.M. (1976). Measurement consequences of selected standard-setting models. *Florida Journal of Educational Research*, 22-27.

Jaeger, R.M. (1989). Certification of student competence. In R. Linn (Ed.), *Educational measurement. Third edition.* New York, NY: American Council on Education and Mac-millan Publishing Company.

Koffler, S.L. (1980). A comparison of approaches for setting proficiency standards. *Journal of Educational Measurement 17*(3), 167-178.

Linn, R. (1978). Demands, cautions, and suggestions for setting standards. *Journal of Educational Measurement 15*(4), 301-308.

Livingston, A.S., & Zieky, M.J. (1989). A comparative study of standard setting methods. *Applied Measurement in Education 2*(2), 121-141.

Lord, F.M (1962). Cutting scores and errors of measurement. *Psychometrika 27*(1), 19-30.

Lorge, I., & Diamond, L.K. (1954). The value of information to good and poor judges of item difficulty. *Educational and Psychological Measurement 14*, 29-33.

Meskauskas, J.A. (1976). Evaluation models for criterion referenced testing: Views regarding mastery and standard setting. *Review of Educational Research 46*(1), 13-158.

Messick, S. (1989). Validity. In R. Linn (Ed.), *Educational measurement. Third edition.* New York, NY: American Council on Education and Macmillan Publishing Company.

Nedelsky, L. (1954). Absolute grading standards for objective tests. *Educational and Psychological Measurement 14*, 3-19.

Popham, J. (1978). As always provocative. *Journal of Educational Measurement 15*(4), 297-300.

Ravelo, N.E., & Nitko, A.J. (1986). Selection bias according to a new model and four previous models using admission data from a Latin American University. ERIC Document 291774.

Shepard, L. (1980). Standard setting issues and methods. *Applied Psychological Measurement 4*(4), 447-467.

Zieky, M.J., & Livingston, S.A. (1977). Manual for setting standards on the basic skills assessment tests. Princeton, NJ: Educational Testing Service.

Chapter 10

Methods and Issues in Setting Performance Standards*

GARY W. PHILLIPS

National Center for Education Statistics, United States Department of Education, Washington DC, USA

A performance standard is the evidence used to indicate the mastery of acceptable levels of student learning. It is often a cut score on a test, but, in general, it can be any demonstration that students have mastered a pre-defined level of learning. This chapter describes various methods of establishing such standards, reviews the technical issues involved in these processes, and provides practical advice on the steps that should be followed. It is emphasized that the setting of student performance standards is ultimately a judgmental activity that should be informed by an empirical process.

A standard is an answer to the question 'How good is good enough'? and this question can only be answered by someone's judgment (Livingston & Zeiky, 1982, p. 12).

The setting of performance standards is a complicated political and scientific activity. The process of setting standards occurs in all areas of physical and social sciences. For example, the speed limit of 110 kilometers per hour set on the motorways in Sweden is a good example of a cut score for a performance standard. How fast automobiles travel is essentially a continuous variable. However, for safety reasons it is practical to establish a speed limit beyond which the society considers that it is unsafe to drive. Similar considerations are taken into account

---

*The views expressed in this chapter are those of the author and do not necessarily represent the position of the National Center for Educational Statistics, United States Department of Education.

when determining the minimum age for voting and for entry into military service. Likewise, environmental scientists have to establish maximum levels of toxic poisons for pesticide use, drinking water, air pollution and radiation exposure. In business, standards for quality and safety are established, and in banking, standards are set on the savings that banks are allowed to invest in bonds. In the end, all of these decisions are arbitrary, but they are almost always informed by a variety of systematic data. Performance standards in education are established with a political sensitivity to social consequences. When policymakers set standards, they try to do so for the public good. However, much of the controversy over standard setting in education is centered around disputes over what is or should be in the best interest of the public.

In education, policymakers often set targets on a variety of education indicators. For example, a ministry of education may decide that a graduation rate of 95 percent is a good target for the secondary school population. This represents a performance standard but it is decided primarily through a political process. The only scientific data that are used in the setting of such standards are usually past measures of the graduation rate and, possibly, data emerging from international comparative studies.

Other education performance standards require significantly more empirical data before a standard can be reasonably set. This is the case when educators attempt to establish acceptable levels of literacy for the adult population, or determine the level of proficiency needed in reading, writing and mathematics in order to be promoted from one grade level to another, or to graduate from secondary school. Considerable empirical data are also needed when educators set standards on tests that are supposed to be predictive of success in tertiary education. These types of performance standards cannot be set solely on the basis of a political process. Although the standards are ultimately judgmental, they involve more than human intuition only. They require a consensus on what is meant by notions such as "literacy", "proficiency", and "success". Once these hypothetical constructs are defined, reliable, valid, and fair measures of them must be developed. Such measures usually take the form of test items that must be aggregated or summarized in some way. In order to set reasonable standards on these summary scores, the standard setters need empirical data that show how the test scores are related to the hypothetical constructs the scores purport to measure, how well the various standards predict desired outcomes or variables external to the test, and how well one can expect students to perform on the test depending on where the standard is established. The continual interaction between informed human judgment and useful empirical data is all part of the art and science of setting performance standards in education.

Performance Standards in the United States

In the United States, almost all past efforts to set performance standards have occurred at the district and State level. This is consistent with the country's decentralized education system, and with the generally accepted notion that the responsibility for establishing the curriculum rests with local and State authorities. Much of the standard setting activities in the 1970s and 1980s involved minimum competency testing (Jaeger, 1989). However, after 20 years of mediocre student performance in the US, as documented by national studies (Mullis, Owen & Phillips, 1990) and international comparisons (McKnight et al., 1987), policymakers begun to realize that in order to compete internationally education standards must be addressed as a national issue. Raising expectations and setting high standards has been embraced as a reasonable role for national policymakers (see also Chapter 1).

In the early 1990s, the idea of setting standards in education reached a zenith in the United States. The major impetus for this activity began with the 1989 Education Summit between the President of the United States and the Nation's Governors. This led to the acceptance of six national education goals in 1990, and the creation of the National Education Goals Panel (NEGP) to monitor progress toward achieving the goals. The six goals adopted were:

1. All children in America will start school ready to learn.
2. The high school graduation rate will increase to at least 90 percent.
3. American students will leave grades four, eight, and 12 having demonstrated competency in challenging subject matter, including English, mathematics, science, history, and geography; and every school in America will ensure that all students learn to use their minds well, so they may be prepared for responsible citizenship, further learning, and productive employment in a modern economy.
4. US students will be first in the world in science and mathematics achievement.
5. Every adult American will be literate and will possess the knowledge and skills necessary to compete in a global economy and exercise the rights and responsibilities of citizenship.
6. Every school in America will be free of drugs and violence and will offer a disciplined environment conducive to learning.

Goal 3 has many of the ingredients of a performance standard. It implies, first, a national consensus on what should be taught and learned in English, mathematics, science, history and geography (*standards for content domains*); second, levels of competency must be established in grades four, eight, and 12 (*performance standards*); and, third, a prediction that reaching the performance standards will result in "responsible citizenship, further learning, and productive employment" (areas to look for *criterion-related validity*).

Shortly after the creation of the NEGP, the National Council on Education Standards and Testing (NCEST) was formed. This subcommittee produced a report, *Raising Standards for American Education* (NCEST, 1992), which concluded that "national standards are desirable" (p. 9). The NCEST panel also distinguished between the "content standards that describe the knowledge, skills and other understandings that schools should teach in order for other students to obtain high levels of competency in challenging subject matter" (p. 13), versus "student performance standards that define various levels of competence in the challenging subject matter set out in the content standard" (p. 13). This distinction between content and performance standards led to a lot of discussion and confusion among policymakers over the definitions of these terms. In 1993, a technical planning group for goals 3 and 4 further clarified the definitions and distinction between content and performance standards:

> Content standards specify what students should know and be able to do. In shorthand, they involve the knowledge and skills essential to a discipline that students are expected to learn. Those "skills" include the ways of thinking, working, communicating, reasoning, and investigating that characterize each discipline. That "knowledge" includes the most important and enduring ideas, concepts, issues, dilemmas, and information of the discipline (NEGP, 1993, p. 9).

At the present time there are many independent efforts to set national content standards in different subject areas. The major efforts along with the group responsible for them include:

Arts—Music Educators National Conference;
Citizenship and Civics—Center for Civic Education;
English, Language Arts—Center for the Study of Reading;
Foreign Languages—American Council on the Teaching of Foreign Languages;
Geography—National Council of Geographic Education;
History—National Center for History in the Schools;
Mathematics—National Council of Teachers of Mathematics;
Science—National Academy of Sciences, National Research Council.

A content standard represents what the students are expected to learn. It often gets operationalized in terms of curriculum guides, becomes the subject of textbooks, and represents the goals and objectives of teaching and assessment. However, a distinction is useful between the range of things students are expected to learn (*content standards*) versus the procedures used to demonstrate the attainment of acceptable levels of learning (*performance standards*).

Performance standards specify 'how good is good enough'. In shorthand, they indicate how adept or competent a student demonstration must be to indicate attainment of the content standards. They involve judgments of what distinguishes an adequate from an outstanding level of performance. Performance standards are not the skills and modes of reasoning referred to in the content standards. Rather, they indicate both the nature of the evidence (such as an essay, mathematical proof, scientific experiment, project, examination, or combination of these) required to demonstrate that content standards have been met and the quality of student performance that will be deemed acceptable (what merits a passing or an *A* grade) (National Education Goals Panel, 1993, p. 22).

The performance standard describes what the student must do in order to demonstrate that acceptable levels of learning have occurred. The performance standard may get operationalized in terms of the number of courses a student should take, or marks on an oral presentation, a demonstration project, or a portfolio. However, in most cases, a performance standard is expressed as a cut score on a test.

In the past, most standard setting activities centered around making decisions about the certification or promotion of individual students. However, in recent years, efforts at setting standards for populations of students have emerged. The main effort to set performance standards at the national level in the United States has been carried out by the National Assessment Governing Board (NAGB).

NAGB has worked for several years to set performance standards (basic, proficient, and advanced) in grades four, eight, and 12 for each of the subjects being assessed in the National Assessment of Educational Progress (NAEP). So far, the Board has established performance standards in mathematics (Mullis et al., 1993a) and reading (Mullis et al., 1993b). The Board tried, but was unable, to set standards in writing. Plans are underway to establish standards in history and geography. The Board used a modified version of Angoff's (1971) procedurem, but has been widely criticized on technical grounds (Livin et al., 1991; Stufflebeam, Jeager & Scriven, 1991; US General Accounting Office, 1993; National Academy of Education, 1993; Burstein et al., 1993).

The US Congress is currently considering legislation proposed by the President that would authorize a host of activities centered around national goals and higher standards. The legislation would add "arts" and "foreign languages" to the subjects included under Goal 3; authorize the NEGP; and create a National Education Standards and Improvement Council (NESIC) to certify voluntary national and state-level content and performance standards. The NESIC would also certify *opportunity to learn standards*—these are the resources and services needed to provide all students with the opportunity to learn the knowledge and skills called for in the content standards.

It is clear from the above that much of the current education reform effort in the United States is centered around the establishment of performance standards.

Standard Setting Procedures

This section deals with the methods and procedures of setting performance standards on education achievement tests. It provides more detailed coverage of some of the procedures that were introduced in the preceeding chapter, although the range of coverage in this chapter is limited to the category of judgmental methods most often used to set standards in education on large-scale criterion-referenced tests (CRT) widely used in district, state, national, or international assessments. A criterion-referenced test is one that is

> ... deliberately constructed so as to yield measurements that are directly interpretable in terms of specified performance standards ... The performance standards are usually specified by defining some domain of tasks that the student should perform. Representative samples of tasks from this domain are organized into a test. Measurements are taken and are used to make a statement about the performance of each individual relative to the domain (Glaser & Nitko, 1971, p. 653).

The above definition has two components. The first is a "criterion"—which is usually an external measure to which the test scores are referenced, or which is predicted by the test scores. It is this criterion that gives the test score its meaning. An example might be a percent correct score that is derived from a sample of items that are representative of a defined domain in algebra. If students receive a 75 percent correct score then this means that they have mastered 75 percent of the content in the domain.

The second part of the definition is the "performance standard". This standard represents the amount of knowledge and skills in the domain that the students should master. The process of determining the performance standard usually consists of making an informed but arbitrary decision about the level of knowledge and skills needed for some purpose, and determining the performance standard on the test that best predicts the desired level of performance on the criterion.

A large number of standard setting methods and three major classifications of procedures are reviewed in Chapter 9. An additional classification makes a distinction between "state models" and "continuum models" (Meskauskas, 1976; Livingston & Zeiky, 1982; Berk, 1986). State models have been of interest to researchers (Macready & Dayton, 1980), but have not received much attention by practitioners. This may be due to the unrealistic assumption that the underlying proficiency is discrete—or binary—rather than continuous. The methods that assume that the test measures a continuous, underlying trait are often classified according to "test-centered" versus "examinee-centered" procedures (Sheppard, 1984; Jaeger, 1989). Although there are dozens of methods (and modifications of methods) that could be covered, this section will only review the most commonly used procedures. This includes the four test-centered approaches by Angoff (1971), Ebel (1972), Nedelsky (1954), and Jaeger (1978; 1982), and three examinee-centered approaches.

Angoff's Method

Angoff's method (1971) is one of the simplest and most widely used methods of setting performance standards on dichotomously scored items. This procedure requires that each judge be asked

> ... to state the probability that the 'minimally acceptable person' would answer each item correctly. In effect, the judges would think of a number of minimally acceptable persons, instead of only one such person, and would estimate the proportion of minimally acceptable persons who would answer each item correctly. The sum of these probabilities, or proportions, would then represent the minimally acceptable score (Angoff, 1971, p. 515).

Although the Angoff procedure is relatively straightforward, there are many suggested modifications to this basic procedure. Both the Ebel and the Jaeger procedures are such modifications.

Ebel's Method

Ebel's (1972) method is a more elaborate form of the Angoff procedure for dichotomously scored items. In this procedure all the items on the test are categorized into a difficulty by relevance matrix. For example, three levels of difficulty might be defined—easy, medium, and hard—usually according to the p-values of the items. Levels of relevance might be—questionable, acceptable, important, and essential. After all items have been categorized into the cells of the difficulty by relevance matrix, the judges are asked to estimate the proportion of items in each cell that "minimally acceptable" persons would answer correctly. The standard for each judge is computed as the weighted average of the proportions in each cell (the proportions in each cell are weighted by the number of items in each cell).

Ebel's and Angoff's methods are similar in that both rely on item judgments and require that judges conceptualize a minimally acceptable examinee. However, the two procedures have several important differences. In Ebel's method, the judges must explicitly rate each item along two dimensions (difficulty and relevance). The theory behind this aspect of the procedure is that this additional information about each item should help the judges make a more informed decision. The second difference is that the standards resulting from Ebel's procedure are based on sets of similar items (or parcels of items) rather than individual items. Many judges may be more familiar and comfortable with making decisions on parcels of items—which are like mini-tests—than individual items.

Nedelsky's Method

The Nedelsky method is the oldest of the test-centered procedures reviewed here. Like Angoff's and Ebel's methods, the Nedelsky approach is an item-based procedure that requires judges to conceptualize a minimally acceptable student. However, it is only appropriate for multiple-choice items. In the Nedelsky procedure the judges are asked to review each item and estimate which options the minimally acceptable student should be able to eliminate as being incorrect. It is assumed that the minimally acceptable student will choose at random from the remaining options implying that the "minimal pass level" for each item will be the reciprocal of the number of remaining options. As an example, if a minimally acceptable student can eliminate two out of four options as incorrect, then the probability of minimally passing the item is $1/(4-2)=$. The standard on the test for each judge is obtained by summing the minimal pass levels across all items.

Nedelsky also provides a methodology—which, incidentally, can also be applied to the Angoff and Ebel procedures—to adjust the initial standard due to mis-classification error resulting from a lack of test reliability. If the classical test theory assumption is made that the errors of measurement are normally distributed and constant throughout the scale, then examinees with true ability equal to the standard will receive scores that are normally distributed with a standard deviation equal to the standard error of measurement. Fifty percent of these students will pass and 50 percent will be mis-classified due to error of measurement even though all of them had true scores that reached the passing standard. This mis-classification rate can be reduced by lowering the standard. For example, if the standard is lowered by one standard error of measurement, the rate of mis-classification for examinees that just barely pass the test drops from 50 percent to 16 percent.

Brennan (1984) and others have questioned the conceptual underpinnings of the Nedelsky method. The sentiment is captured by Shepard (1984) when she states "the method is conceptuality flawed because there is no good reason or evidence to support the idea that examinees that know the answer guess at random" (p. 177). She also argues that standards set with the Nedelsky method tend to be lower than those set using other methods (Shepard, 1980).

Jaeger's Method

Jaeger's method is another variant of the Angoff procedure. Jaeger asks judges to review each item and make a yes/no decision as to whether every examinee in the test population who meet the standard should correctly answer this item. Although Jaeger does not ask judges to imagine a minimally acceptable examinee, he does require judges to imagine an acceptable examinee. Jaeger also recommends that judges be given empirical data about the consequences of their standards and be given several opportunities to revise their standards in the light of this additional information.

Several researchers have discovered problems with the Jaeger method. Beck (1986) argued that binary choices of *0 and 1* are less realistic than probability choices that range from *0 to 1*. This same complaint was given as the reason Cross et al. (1984) found Jaeger's method less reliable than Angoff's.

Discussion

In general, there are several problems associated with previously developed test centered approaches to standard-setting. One problem is that they typically deal with dichotomously scored items and do not accommodate polychotomous rating scales. Yet, many of the more recent reforms in assessment require complex ratings. Performance assessments using essays, demonstration projects, portfolios, and other forms of "authentic" testing, are not easily adaptable to the test-centered approach to standard setting.

A second problem is that the above methods usually require that probability values be summed across a large number of items to enhance the stability of the performance standard. Yet many recent large scale assessments use very few performance type items because they take so much time to administer and cost so much to score. Things are even more complicated when the same assessment includes both multiple-choice and performance type items. The recent experience of the National Assessment Governing Board at setting performance standards on the NAEP found that the standards set by constructed response items were substantially higher than the standards set by multiple-choice items.

A third difficulty with the above approaches is that they require a undimensional set of items. Again, this assumption is being challenged by performance assessments, which are often so explicitly multi-dimensional that they are separately scored at the individual exercise level. This would literally require that a performance standard be established for each individual exercise.

A final difficulty with item-based approaches is that the cognitive complexities associated with setting standards across so many items may be overwhelming for judges. In the recent evaluation of the achievement levels set by the NAGB, the level setting process was deemed to be "fundamentally flawed". This conclusion was reached because it was felt that such item-based approaches presented "an unreasonable cognitive task because judges have no basis for making such judgments. They nonetheless respond to the demand characteristics of situation but in idiosyncratic ways, relying ... on personal experience, opinion, and intuition" (National Academy of Education, 1993, p. 72). More research is needed to further investigate this argument.

Examinee-Centered Approaches

The above procedures are based on information internal to the test. Examinee-centered approaches use information external to the test as a way of classifying examinees. The test standard is then determined through the relationship of the test to the classification of the examinees. Since the examinee-centered approaches use data external to the test they are often used as validation studies for standards derived from test centered approaches. Several of these approaches will be briefly described.

In the *borderline method* (Zeiky & Livingston, 1977) a group of judges classify examinees into three categories—competent, borderline, and incompetent—based on auxiliary information that is not related to the test (e.g., grade point average, scores on other tests, teachers observations, etc.). The test scores of the borderline group are then analyzed, and the median is taken as the standard on the test. The *contrasting-groups method* (Zeiky & Livingston, 1977) follows the same procedures as the borderline method. The difference is in the analysis of the scores on the test. Instead of taking the median of the borderline group as the standard, the contrasting-group method uses one of several techniques to distinguish the competent from the incompetent examinees. Livingston and Zeiky (1982) provide a variety of procedures for accomplishing this task.

Another examinee centered approach is called the *bootstrapping method* (Glass, 1978). This procedure is carried out by correlating the CRT with other criteria and choosing the point on the CRT-scale that predicts competence or success on the criterion. For example, a school district may establish a performance standard on a diagnostic test in reading that predicts success in passing a state-mandated reading test used for promotion or graduation. A recent variation of this procedure is referred to as the *benchmarking approach* (Resnick & Nolan, 1994). In benchmarking, the standard setter attempts to find examples of best practice or high performance as an aid in setting the standard. For example, on international studies, the United States may look at the high mathematics performance of Dutch, Korean or Japanese students and set a standard that reflects similar high performance in the United States. Alternatively, one can examine the performance levels of students of high-performing states such as Iowa, North Dakota, and Minesota (NCES, 1993). A recent example of benchmarking was carried out in a research study by Pashley and Phillips (1993). The authors equated the scores on the NAEP to the scores on the second International Assessment of Educational Progress (Lapointe, Mead & Askew, 1992). The study permitted the NAGB to evaluate how well American students performed on a US standard when benchmarked to an international context.

Setting Standards through Direct Judgment

The above methods are commonly used statistical approaches. They are statistical in the sense that human judgments are quantified and then subjected to a statistical process that ultimately determines the performance standard. However, there are other less statistical procedures that have been used to set standards on achievement tests. These methods usually involve the direct judgment of one, or, a vote among several persons. As such, they are not so much methods of setting standards as they are methods of declaring standards by decree.

One approach is to choose a performance standard by selecting conventional points along a norm-referenced scale using standard deviation units, percentiles or quartiles. This approach makes a lot of sense, for example, if the top 10 percent of students are to be selected for program placement. The standard has more to do with how many students can be accommodated in the program, and less with what they know and can do. However, in some cases this approach can also have criterion-referenced interpretations. For example, since 1984 the NAEP has used a technology called *behavioral anchoring* to give criterion-referenced meaning to the standard deviation units on the (0-500) NAEP scale (levels 200, 250, 300, and 350). For example, in the 1992 mathematics assessment (Mullis et al., 1993a) the students are able to do the following tasks at each standard deviation unit:

> 200 – Addition, subtraction, and simple problem solving with whole numbers;
> 250 – Multiplication, division, simple measurement, and two-step problem solving;
> 300 – Reasoning and problem solving, fractions, decimals, percents, elementary concepts in geometry, statistics and algebra;
> 350 – Reasoning and problem solving involving geometric relationships, algebra and functions.

Although the behavioral anchoring technology has been mostly applied to standard deviation units, it can theoretically be applied to percentiles and quartiles, or to any other standard established on the scale.

A second direct judgment method to set performance standards is to establish them as part of a *scoring rubric*. This is particularly useful when the CRT is composed of performance type tasks. For example, many direct assessments of writing involve scoring protocols which define levels of writing proficiency such as unsatisfactory, minimal, adequate, and advanced. One of these definitions may be treated as a standard at the item level (such as considering "adequate" as the performance standard), and when aggregated, represent standards at the total test level. A similar approach may be applied to other types of performance assessments such as essays, portfolios, demonstrations, and projects.

Comparability of Standard Setting Methods

The most consistent finding in the standard setting literature is that different standard setting methods yield inconsistent results (Glass, 1978).

The inconsistencies appear to be artifacts of the methods employed (Poggio, Glasnapp & Eros, 1981) and the categories of judges used (Jaeger, Cole, Irwin & Pratto, 1980). In a review of the research literature, Jaeger (1989) summarized the results of 32 comparisons culled from 12 separate studies. The results of the studies indicated that, on the average, almost six times as many students would fail when using one standard setting method rather than another. He concluded that "there is little consistency in the results of applying different standard-setting methods under seemingly identical conditions, and there is even less consistency in the comparability of methods across settings" (Jeager, 1989, p. 500). Jaeger goes on to say that Hambleton (1980), Koffler (1980), Sheppard (1980) and he himself (Jeager et al., 1980; Jeager & Busch, 1984) recommend that "it might be prudent to use several methods in any given study than consider all of the results, together with extra statistical factors, when determining a final cut score" (p. 500). This lack of comparability across different standard setting methods led other reviewers to conclude that standard setting is the Achilles' heel of criterion-referenced testing (Sheppard, 1984, p. 169). Even more pessimistic is the review by Glass (1978), who argued:

> A common expression of wishful thinking is to base a grand scheme on a fundamental, unsolved problem ... Those who think on exalted levels are prone to underrate the complexity of what seem lesser problems ... I have read the writing of those who claim the ability to make the determination of mastery or competence in statistical or psychological ways. They can't (p. 237).

Glass (1978) subsequently reiterates that "every attempt to derive a criterion score is either blatantly arbitrary or derives from a set of arbitrary premises" (p. 258). Others have disagreed with Glass (e.g., see Popham, 1978a), arguing that although standard setting may be arbitrary, the process need not be capricious. There are many situations in education, business, and psychology in which continuous variables must be artificially categorized. Since standard setting is often dictated by practical necessity, it should be done in as thoughtful a way as possible.

As the above reviews indicate, it is widely accepted that there is a considerable method effect among standard setting procedures. Practitioners should recognize this fact and realize that standard setting is always a judgmental process that, at best, can be informed by (possibly multiple forms of) empirical standard setting procedures.

Reliability of Scores based on Performance Standards

In general, the reliability of a test "refers to the degree to which test scores are free from errors of measurement ... Fundamental to the proper evaluation of a test are identification of major sources of measurement error, the size of the errors resulting from these sources, the indication of degree of reliability to be expected between pairs of scores under particular circumstances, and the generalizability of results across items, forms, rates, administrations, and other measurement facets" (AERA, 1985, p. 19). In addition to demonstrating that the test itself is reliable for a particular use, test developers should also demonstrate that the examinees will be consistently classified based on performance standards established on the test. This is often operationalized as providing an index of the "percentage of test takers who are classified in the same way on two occasions or on alternate forms of the test" (AERA, 1985, p. 23). Classification consistency is especially important on tests designed to make decisions about individual students. This is because errors in classification may harm students by, for example, preventing deserving students from graduating, or promoting students who are unprepared for the next grade level. Classification error associated with group level assessments are less important because errors of classification tend to cancel out in the aggregate.

Binomial Error Model

Several statistical procedures have been developed to estimate the classification consistency on tests with performance standards using the binomial error model. The first to use the *binomial error model* was Subkoviak (1976). In this model it is assumed that the examinee's conditional observed test scores are independent with a binomial distribution. The parameters of the binomial distribution are number of test items (n) and the examinees' true score (θ). For each examinee, with true proficiency θ, the probability of reaching the performance standard can be found from the tables of the binomial distribution. Subkoviak (1976) recommends that θ be estimated by:

$$\hat{\theta} = \rho_{xx}(x/n) + (1-\rho_{xx})\bar{x}/n \qquad \text{Eqn 10.1}$$

Where x equals the examinees raw score, x equals the average raw score for the population of examinees, and ρ_{xx} equals the classical test theory reliability of the test using either the KR-21 method (Subkoviak, 1976) or the KR-20 method (Subkoviak, 1984). The probability that an individual examinee will be declared a master, $Px=\text{prob}(x \geq \theta)$, is determined by using tables for the cumulative binomial distribution. The probability that an individual examinee will be consistently classified on two equivalent forms of the same test is found by $Px^2+(1-Px)^2 = 1-2(Px-Px^2)$. If $f(x)$ is the density of x, the proportion of examinees in the population who would be consistently classified is:

$$\hat{p} = \sum f(x)[1-2(Px-Px^2)] \qquad \text{Eqn 10.2}$$

The proportion of examinees one would expect to be consistently classified by chance is found by:

$$\hat{P}_c = 1-2\{\ \Sigma f(x)Px - [\Sigma f(x)Px]^2\}$$ Eqn 10.3

The proportion of examinees which would be consistently classified, adjusted by chance, can be estimated using Cohen's (1960) coefficient Kappa:

$$\hat{\kappa} = (\hat{p}-\hat{P}_c)/(1-\hat{P}_c)$$ Eqn 10.4

Additional Error Models

Another version of the binomial error model is the procedure recommended by Huynh (1976), which uses the *beta-binomial error model* (Keats & Lord, 1962). Huynh's approach leads to estimates of p and K, by assuming that the population of true scores are distributed as a beta distribution, and the conditional distribution of observed scores is a beta-binomial distribution.

Another set of reliability coefficients frequently used in standard setting are based on extensions of traditional classical test theory models. For example, Livingston (1972) estimated a reliability coefficient by generalizing the KR-20 index. This was accomplished by calculating deviations from the performance standard rather than the mean. The Livingston index, k^2, is found by:

$$k^2 = \frac{\sigma_\tau^2 + (\mu_x-x_c)^2}{\sigma_\tau^2 + \sigma_e^2 + (\mu_x-x_c)^2}$$ Eqn 10.5

where σ_τ^2 equals the variance of true scores, σ_c^2 equals the error variance, μ_x equals the mean of the population, and x_c equals the performance standard. When $\mu_x=x_c$ the Livingston index equals KR-20. Some very sophisticated extensions of the Livingston index have been provided by Brennan and Kane (1977) using an application of generalizability theory to criterion-referenced tests.

Validity of Performance Standards

Messick (1989) defines validity as an

... integrated evaluative judgment of the degree to which empirical evidence and theoretical rationales support the adequacy and appropriateness of inferences and actions based on test scores or other modes of assessment ... Hence, what is to be validated is not the test or the observation device as such but the inferences derived from test scores or other indicators ... It is important to note that validity is a matter of degree not all or none. Furthermore, over time, the existing validity evidence becomes enhanced (or contravened) by new findings ... Inevitably, then, validity is an evolving property and validation is a continuing process (p. 13).

Although Messick's definition was intended for test scores it applies equally well to performance standards established on test scores.

Conventional wisdom focuses on three forms of validity evidence, content, construct and criterion-related validity (e.g., see AERA, 1985). As noted in Chapter 8, *content validity* refers to evidence demonstrating how well the test is representative of the content domain. Most content validity studies are carried out by curriculum committees that independently verify that the test items are a representative sample of the content domain. Cronbach (1971) provided a statistic for content validity by examining the equivalence of two tests developed from the same content domain by independent groups of test developers.

Construct validity refers to evidence demonstrating the extent to which a test measures a hypothetical construct—such as algebra proficiency.

It should be noted that establishing the validity of a measure of a construct is a problem distinct from that of using that measure in predicting a second measure, although the latter can often contribute to construct validation. The construct of interest for a particular test should always be embedded in a conceptual framework, no matter how imperfect that framework may be. The conceptual framework specifies the meaning of the constructs, and indicates how measures of the construct should relate to other variable (AERA, 1985, pp. 9-10).

Although content and construct validity are important issues for criterion-referenced tests, it is criterion-related validity evidence that is especially important when performance standards are established. CRTs are developed because we want an indication of how much a student has learned with reference to some criterion. The validity or usefulness of the criterion-referenced test is normally evaluated in terms of its accuracy at predicting criterion performance. For example, the coefficient of determination, r^2_{yx}, is an index of how much variance in the criterion, y, can be can be predicted from the test, x. As was the case with reliability, there is a distinction between validity of the inferences based on the test, versus the validity of inferences based on the standard on the test.

When a standard is established on a CRT, it is supposed to be related to the level of acceptable proficiency on a criterion. The prediction problem is one of predicting a discrete, rather than a continuous, outcome. A concrete example might help to clarify concepts and facilitate further discussion. Suppose that a CRT has been developed to be representative of the domain of algebra explicated in a 10th grade curriculum framework. The framework outlines the scope and sequence of algebra instruction recommended for the population of students and represents the types and levels of knowledge and skills students should know by the time they complete the 10th grade. The curriculum framework is the "criterion" to which the CRT is to be referenced. In the United States this curriculum framework has recently been referred to as the *content standard.* A definition of an acceptable level of proficiency is established by curriculum experts and policymakers as an aid to establishing the *performance standard.*

It should be noted that both the criterion and the acceptable levels of proficiency are external and possibly developed prior to the test. Suppose moreover that the definition of acceptable levels of proficiency is used to establish a cut score of 75 percent in the algebra CRT (possibly through one of the six methods described above). The cut score of 75 percent is purported to be predictive of the acceptable levels of proficiency on the criterion. If a representative sample of 10th grade algebra students is taken and their algebra teachers are asked to place their students in two categories—those that have clearly not reached the performance standard and those that clearly have, then a discrete external criterion variable is established that one wants to be able to predict from the cut score. This situation results is a four-way classification matrix as shown in Table 10.1.

Table 10.1. Four-way classification matrix

Performance criteria	Cut-score	
	Below	Above
Acceptable	False negative	Positive
Unacceptable	Negative	False postive

The four cells of the table can be described as follows:

1. Positive—the proportion of examinees who reached the cut score on the CRT and also met the performance standard based on the external criterion;
2. Negative—the proportion of examinees who failed to reach the cut score on the CRT and also failed to meet the performance standard based on the external criterion;
3. False Positive—the proportion of examinees who reached the cut score on the CRT but failed to reach the performance standard based on the external criterion; and

4. False Negative—the proportion of examinees who failed to reach the cut score on the CRT but met the performance standard based on the external criterion.

How well the cut score predicts or reflects the performance standard is the essential criterion-related validity question, and depends on three things: the correlation between the CRT and the criterion; the base rate, or proportion of students who meet the performance standard in the population; and, how high or low the cut score is set on the test.

A number of researchers have used a variety of correlational and decision theory-models to provide statistical indices of criterion-related validity. For example, Taylor and Russell (1939) estimate the proportion of positive decisions for various cut scores, given a fixed correlation and base rate, and Brogden (1946) reported similar estimates for the percent of correct decisions (both positive and negative). Validity coefficients can also be calculated within the context of aptitude-by-treatment interactions when the CRT is used for purposes of instructional placement or classifications (Cronbach, 1971). Evidence of validity is obtained by regressing the criterion measure on the CRT and finding significant differences in the regression coefficients between the placements or classifications. Other approaches use some variant of the contrasting groups approach as a method of validation rather than a method of setting standards. For example, Hambleton (1984) suggests that performance standards based on an external criterion be obtained from an instructed group (e.g., the students who clearly have the competencies to graduate from high school) and an uninstructed group (e.g., the students who clearly do not have the competencies required to graduate from high school). The above four-way classification table is then formed and the validity coefficient is the proportion of correct decisions. Other refinements may be found in Berk (1976), Popham (1978b), Shepard (1984), Kane (1982), and Millman (1979).

Although the above three approaches represent the historically traditional thinking about validity, recent views have emphasized a more consequential basis for validity (Messick, 1989). The argument is that one must go beyond the traditional view of validity that deals primarily with evaluating how well the intended outcomes of testing are accomplished to a concern with the social consequences of testing.

> The point is that the functional worth of testing depends not only on the degree to which the intended purposes are served but also on the consequences of the outcomes produced ... especially those un-intended side effects that are remote from the expressed testing aims (Messick, 1989, p. 85).

This line of thinking is especially relevant to assessments that use complex performance-type tasks. For example, multiple-choice test questions can be used to achieve high degrees of traditional reliability and validity in assessing reading, writing and math skills. However, an unintended educational consequence is that this may lead to an instructional emphasis on basic skills and memorization. On the other hand, it may be difficult to achieve high levels of traditional reliability and validity using performance tasks, essay projects, and portfolios. Yet these types of assessments often lead teachers to emphasize the kind of higher order thinking skills that most educators would prefer. Most educators would prefer this latter outcome.

The consequential basis to validity applies not only to the test but also to performance standards established on the test. This is why it is highly recommended that standard setters receive as much feedback as possible regarding the consequences of their standards before they make any final decisions. The judges should receive distributional information for the general population as well as for all relevant subgroups. This permits the judges to evaluate the extent to which the performance standard is fair for all students.

Implementing Standard Setting Approaches

The above discussion represents the essential ingredients of the most commonly employed methods of setting standards. However, their successful implementation often involves many additional activities in actual practice. The following is a set of recommendations that, if followed, should improve the usefulness of the results from any of these procedures.

Regardless of the procedures chosen, the practioner should follow at least the following steps:

1. Provide clear definitions of the knowledge and skills expected of minimally acceptable examinees.
2. Make sure that the knowledge and skills expected are actually assessed on the test.
3. In most cases multiple judges will be used. When this is the case, then determine the target population of judges that are needed to set the standards, and use appropriate sampling methods that guarantee that the sample of judges is representative of the population.
4. Train the judges to a level in which they understand and are comfortable with the definitions of the minimally acceptable person, the content domain of the test, and the items on the test.
5. Provide the judges with empirical feedback about the consequences of their decisions. The empirical feedback might take the form of distributional information, or pass/fail rates by gender, race, etc. Such feedback helps insure that the judges are not "flying without radar" when they make their decisions.
6. Give the judges a chance to revise their ratings in the light of the empirical consequences. This iterative process might be repeated several times with more or different types of feedback. Such a process would allow the judges to determine if their ratings are too high, or low, or too variable.

7. Field test the whole standard setting process so that all technical and administrative procedures can be worked out ahead of time.

8. Allow a period of review and comment from various stakeholders. The reviews and comments are really another form of feedback that help the standard setters to better understand and evaluate the consequences of their actions.

9. Use at least two methods of setting standards (preferably one test-centered and one examinee-centered) as a way to examine the effect of the method used in the standard setting.

10. Calculate a measure of judge inconsistency (i.e., the standard error of the standard). This statistic will help evaluate the degree of consensus reached on the standard by the judges. If the judges inter-consistency is very low, then consider the possibility of raising or lowering the standard by one or two standard errors. This procedure reduces classification errors for individual students. However, raising or lowering the standard in this way usually makes less sense for group assessments because the classification errors are assumed to cancel out at the group level.

11. Provide evidence of the reliability of the standard, that is, the degree to which the examinees are consistently classified across two independent administrations of the same test. As mentioned above, this is especially important when standards are used to make decisions for individual students because of the potentially harmful effects of classification error. For group assessments this is less important.

12. Provide evidence that all inferences made from the standard are valid. This implies that all such inferences have to be clearly specified and evidence is collected that supports the correctness of these inferences. This usually involves providing evidence of content, construct and criterion-related validity. In addition, evaluate the consequential efforts of the performance standard by assessing the intended and unintended educational and social impact of the standard.

13. Provide evidence that the performance standards are fair for all subgroups of the population. This includes minority students, and those with handicapping conditions and limited language proficiency and other relevant groups that may be adversely affected by the standards.

Conclusion

Setting performance standards is a little like cooking soup. The recipe calls for a delicious blend of art and science. Experience and experimentation will improve the flavor, which is influenced by tradition and cultural preferences. However, no matter how well the recipe is followed, and no matter how pleased the cook is with the outcome, there will be those who do not like the taste. Public policies often require that standards be established out of practical necessity even though the activity does not have a firm scientific basis. Standard setting is best accomplished in an environment in which policymakers with a vision of the public good work collaboratively with technicians willing to take a risk.

References

AERA/APA/NCME Joint Committee (1985). *Joint technical standards for educational and psychological testing*. Washington, DC: American Psychological Association.

Angoff, W.H. (1971). Scales, norms, and equivalent scores. In R.L. Thorndike (Ed.), *Educational Measurement* (2nd edn, pp. 508-600). Washington, DC: American Council on Education.

Berk, R.A. (1976). Determination of optimal cutting scores in criterion-referenced measurement. *Journal of Experimental Education 45*, 4-9.

Berk, R.A. (1986). A consumer's guide to setting performance standards on criterion-referenced tests. *Review of Educational Research 56*, 137-172.

Brennan, R.L. (1984). Estimating the dependability of the scores. In R.A. Berk (Ed.), *A guide to criterion-referenced test construction* (pp. 292-334). Baltimore, MD: John Hopkins University Press.

Brennan, R.L., & Kane. M.T. (1977). An index of dependability for mastery tests. *Journal of Educational Measurement 14*, 277-289.

Brogden, H.E. (1946). On the interpretation of the correlation coefficient as a measure of predictive efficiency. *Journal of Educational Psychology 37*, 65-76.

Burstein, L., Koretz, D.M., Linn, R.L., Sugrue, B., Novak, J., Lewis. E., & Baker, E. (1993). *The validity of the 1992 NAEP achievement level descriptions as characterizations of mathematics performance.* Los Angeles, CA: Center for Research on Evaluation, Standards, and Student Testing, 1993.

Cohen, J. (1960). A coefficient of agreement for nominal scales. *Educational and Psychological Measurement 20*, 37-46.

Cronbach, L.J. (1971). Test validation. In R.L. Thorndike (Ed.), *Educational measurement* (2nd edn, pp. 443-507). Washington, DC: American Council on Education.

Cross, L.H., Impara, J.C., Frary, R.B., & Jaeger, R.M. (1984). A comparison of three methods for establishing minimum standards on the National Teachers Examinations. *Journal of Educational Measurement 21*, 113-130.

Ebel, R.L. (1972). *Essentials of educational measurement*. Englewood Cliffs, NJ: Princeton-Hall.

Glaser, R., & Nitko, A.J. (1971). Measurement in learning and instruction. In R.L. Thorndike (Ed.), *Educational measurement* (2nd edn). Washington, DC: American Council on Education.

Glass, G.V. (1978). Standards and criteria. *Journal of Educational Measurement 15*, 237-267.

Hambleton, R.K. (1980). Test score validity and standard-setting methods. In R.A. Bede (Ed.), *Criterion-referenced measurement: The state of the art* (pp. 80-123). Baltimore, MD: John Hopkins University Press.

Hambleton, R.K. (1984). Validating test scores. In R.A. Berk (Ed.), *A guide to criterion-referenced test construction* (pp. 199-230). Baltimore, MD: John Hopkins University Press.

Huynh, H. (1976). On the reliability of decisions in domain-referenced testing. *Journal of Educational Measurement 13*, 253-264.

Jaeger, R.M. (1978). A proposal for setting a standard on the North Carolina high school competency test. Paper presented at the meeting of the North Carolina

Association for Research in Education, Chapel Hill, NC. Greensboro, NC: University of North Carolina at Greensboro, Center for Educational Research and Evaluation.

Jaeger, R.M. (1982). An interative structural judgment process for establishing standards on competency tests: Theory and application. *Educational Evaluation and Policy Analysis 4*, 461-476.

Jaeger, R.M. (1989). Certification of student competence. In R.L. Linn (Ed.), *Educational measurement* (3rd edn, pp. 485-514). New York: Macmillan.

Jaeger, R.M., & Busch, J.C. (1984). *A validation and standard-setting study of the General Knowledge and Communications Skills Test of the National Teachers Examination. Final report.* Greensboro, NC: University of North Carolina at Greensboro, Center for Educational Research and Evaluation.

Jaeger, R.M., Cole, J., Irwin, D., & Pratto, D. (1980). *An iterative structured judgment process for setting passing scores on competency tests: Applied to the North Carolina high school competency tests in reading and mathematics.* Greensboro, NC: University of North Carolina at Greensboro, Center for Educational Research and Evaluation.

Kane, M.T. (1982). The validity of licensure examinations. *American Psychologist 37*, 911-918.

Keats, J.A., & Lord, F.M. (1962). A theoretical distribution for mental test scores. *Psychometrika 27*, 59-72.

Koffler, S.L. (1980). A comparison of approaches for setting proficiency standards. *Journal of Educational Measurement 17*, 167-178.

Lapointe, A.E., Mead, N.A., & Askew, J.M. (1992). *Learning mathematics.* Princeton: Educational Testing Service.

Levin, R.L., Koretz, D.M., Baker, E.L., & Burstein, L. (1991). *The validity and creditability of the achievement levels for the 1990 National Assessment of Educational Progress in mathematics.* Los Angeles, CA: Center for Research on Evaluation, Standards, and Student Testing.

Livingston, S.A. (1972). Criterion-referenced applications of classical test theory. *Journal of Educational Measurement 9*, 13-26.

Livingston, S.A., & Zeiky, M.J. (1982). *Passing scores: A manual for setting standards of performance on educational and occupational tests.* Princeton, NJ: Educational Testing Service.

Macready, G.B., & Dayton, C.M. (1980). The nature and use of state mastery models. *Applied Psychological Measurement 4*, 493-516.

McKnight, C.C., Crosswhite, F.J., Dossey, J.A., Kifer, E., Swafford, J.O., Travers, K.J., Cooney, T.J. (1987). *The underachieving curriculum: Assessing US school mathematics from an international perspective.* Champaign, Illinois: Stipes Publishing Company.

Meskauskas, J.A. (1976). Evaluation models for criterion-referenced testing: Views regarding mastery and standard setting. *Review of Educational Research 45*, 133-158.

Messick, S. (1989). Validity. In R.L. Linn (Ed)., *Educational measurement* (3rd edn, pp. 13-104). New York: Mcmillan.

Millman, J. (1979). Reliability and validity of criterion-referenced test scores. In R.E. Traub (Ed.), *New directions for testing and measurement (No. 4): Methodological developments* (pp.75-92). San Francisco, CA: Jossey-Bass.

Mullis, I.V.S., Owen, E.H., & Phillips, G.W. (1990). *America's challenge: Accelerating academic achievement. A summary of findings from 20 years of NAEP.* Washington, DC: US Government Printing Office.

Mullis, I.V.S., Dossey, J.A., Owen, E.H., & Phillips, G.W. (1993a). *The 1992 mathematics report card for the nation and the states.* Washington, DC: US Department of Education.

Mullis, I.V.S., Campbell, J.R., & Farstrup, A.E. (1993b). *The 1992 reading report card for the nation and the states.* Washington, DC: US Department of Education.

Nedelsky, L. (1954). Absolute grading standards for objective tests. *Educational and Psychological Measurement 14*, 3-19.

Pashley, P.J., & Phillips, G.W. (1993). *Toward world-class standards: A research study linking international and national assessments.* Princeton: Educational Testing Service.

Poggio, J.P., Glassnapp, D.R., & Eros, D.S. (1984, April). An empirical investigation of the Angoff, Ebel, and Nedelsky standard setting methods. Paper presented at the meeting of the American Educational Research Association, Los Angeles.

Popham, W.J. (1978a). As always, provocative. *Journal of Educational Measurement 15*, 297-300.

Popham, W.J. (1978b). *Criterion-referenced measurement.* Englewood Cliffs, NJ: Prentice-Hall.

Resnick, L.B., & Nolan, K.J. (1994). Benchmarking education standards. Paper prepared for INES Network A on student outcomes, April 1994. Paris: OECD, Center for Educational Research and Innovation.

Shepard, L.A. (1980). Technical issues in minimum competency testing. In D.C. Berlinger (Ed.), *Review of research in education, Vol. 8* (pp. 30-82). Washington, DC: American Educational Research Association.

Shepard, L.A. (1984). Setting performance standards. In R.A. Beck (Ed.), *A guide to criterion-referenced test construction* (pp. 169-198). Baltimore, MD: John Hopkins University Press.

Stufflebeam, D., Jaeger, R.M., & Scriven, M. (1991). *Summative evaluation of the National Assessment Governing Board's inaugural 1990-91 effort to set achievement levels on the National Assessment of Educational Progress.* Washington, DC: National Assessment Governing Board.

Subkoviak, M.J. (1976). Estimating the reliability from a single administration of a criterion-referenced test. *Journal of Educational Measurement 13*, 265-276.

Subkoviak, M.J. (1984). Estimating the reliability of mastery-non-mastery classifications. In R.A. Berk (Ed.), *A guide to criterion-referenced test construction* (pp. 267-291). Baltimore, MD: John Hopkins University Press.

Taylor, H.C., & Russell. J.T. (1939). The relationship of validity coefficients to the practical effectiveness of tests in selection. *Journal of Applied Psychology 23*, 565-578.

US, General Accounting Office (1993). *Educational achievement standards: NAGB's approach yields misleading interpretations.* (Report No.GAO/PEMD-93-12). Washington, DC: Government Printing Office.

US, National Academy of Education (1993). *Setting performance standards for student achievement. A report of the National Academy of Education Panel on the evaluation of the NAEP trial state assessment: An evaluation of the 1992 achievement levels.* Stanford, CA: The National Academy of Education.

US, National Center for Education Statistics (1993). *Education in states and nations: Indicators comparing U.S. states with the OECD countries in 1988.* Washington, D.C.: National Center for Education Statistics, US Department of Education.

US, National Council on Education Standards and Testing (1992). *Raising standards for American education.* Washington, DC: National Council on Education Standards and Testing.

US, National Education Goals Panel (1991). *The national education goals report: Building a nation of learners. 1991.* Washington, DC: US Government Printing Office.

US, National Education Goals Panel (1993). *Promises to keep: Creating high standards for American students.* Washington, DC: Goals 3 and 4 Technical Planning Group on the Review of Education Standards, National Education Goals Panel.

Zeiky, M.J., & Livingston, S.A. (1977). *Manual for setting standards on basic skills assessment tests.* Princeton, NJ: Educational Testing Service.

Chapter 11

Assessment of Educational Change: A Review of Selected Threats to Validity

MARY LYN BOURQUE

National Assessment Governing Board, Washington DC, USA

This chapter draws heavily from the US experience of the 25-year history of the National Assessment of Educational Progress in examining the measurement of educational change. Measuring change is fraught with threats to both internal and external validity. This chapter examines some of the more common threats, as well as the important elements of assessment design which impact on identifying change: the use of linking exercises for stability in trend measures, accomodating changes in the content to changes in the measure, and the use of adjusted scores for reporting results. Examples are provided for how one large-scale assessment system has capitalized on current technologies in assessment design, data analysis, and reporting mechanisms to ensure valid and reliable measurement of trend.

The National Assessment of Educational Progress (NAEP) is a congressionally-mandated monitoring system in the United States. It is designed to provide an on-going measure of the academic performance of American students in selected grades and at selected ages. Since the first assessment cycle was conducted in 1969, NAEP has evolved in many ways, in part as a result of advances in the field of measurement, and in part as a response to changes in the socio-educational climate of the country. The report, *A Nation at Risk* (National Commission on Excellence in Education, 1983) was a wake-up call to educators, policymakers, parents and employers. As mentioned in Chapter 1, it called attention to the position that something was radically wrong with the education system in the United States. It was claimed that drop out rates were increasing year after year, that young children were starting their experience of formal education less prepared than before, and that students graduating from the nation's secondary schools were less able to achieve adequately in further education, military service, and the workplace.

Despite this dismal message from a decade ago, change has proved difficult to attain. By 1988 the Congress of the United States had authorized the expansion of NAEP to include state representative samples from volunteering jurisdictions interested in measuring the academic achievement of their students. The following year the nation's Governors collaborated with the Bush administration to put forth six national goals. It was expected that these national goals, which are described in full in the previous Chapter, would drive educational reform throughout the rest of the 20th century and into the next (National Governors Association, 1991).

Two of those goals, *academic competency over challenging subject matter*, and *world-class performance in mathematics and science*, could be monitored by the NAEP provided that the National Assessment adopted a standards-based reporting format. In addition to describing what students know and can do, NAEP could also—at least in principle—provide answers to the question, *how good is good enough?*

The purpose of this chapter, therefore, is to discuss some of the common threats to the validity of standards-based reporting of assessment data in general, and trend data in particular. The NAEP is one example of an assessment program that employs cross-sectional data sets to measure change in academic achievement. The fact that NAEP is not longitudinal in nature requires a very complex design in order to minimize such threats to validity and to exploit the data gathered so that the inferences drawn will be valid and reliable. Many of the technical issues encountered have no simple answers, and often the choices are trade-offs between which set of problems one is willing to live with. This chapter is an effort to explore such choices.

Standards for Interpreting National Assessment Data

In the very earliest mode, NAEP data were reported using item difficulty values. These had some utility because they offered content-specific information. If examined holistically, they could also be seen as estimates of the overall level of student achievement. Item difficulty values measured as the percentage of the examinees who respond correctly to the items continue to be reported in NAEP to this day. However, as item pools grew larger, and the number of academic subjects to be included in the assessments proliferated, the reporting of measures of item difficulty became unwieldy. More importantly, such measures had only limited utility for policymakers. The users of the NAEP data were interested in distributional data: statistics that would describe aspects of student performance across the full range of achievement, and that would allow comparisons to be made among various subpopulations.

The NAEP search for summary statistics which could capture the performance of American students in a more meaningful way eventually led to a variety of reporting options, including average percent correct, and ultimately the NAEP scale (Messick et al., 1983; Phillips et al., 1993).

The NAEP scale places the performance of examinees in grades 4, 8, and 12 on a single, unidimensional scale. However, unlike other widely used scales—such as the Scholastic Aptitude Test (SAT) scale or the grading scale *(A-B-C-D-F)*, which have intuitive meaning to a broad audience—it seemed necessary in NAEP to find a way to describe *what students know and can do* in terms of the curriculum *content* the test assessed. This content had, moreover, to be placed on a scale with values ranging from 1 to 500. The method chosen for accomplishing this became known as *scale anchoring*. This method links selected points on the scale to the empirical performance of examinees at and around those points in such a way that it results in a description of performance (called *anchor levels*). Since 1983, scales have been developed in each of the NAEP content areas, and in most areas, subscales representing significant elements of the content domain have been developed as well.

Figure 11.1 displays the 1992 Grade 4 mathematics items mapped onto the NAEP scale. The scale value for each item represents the point at which the examinees performing at that level have an 80 percent probability of responding correctly to the item. Note that the mean score for fourth graders was 218, and that virtually no examinees performed above 300 on the scale.

The anchor points selected varied somewhat across content areas, but commonly used were 200, 250, 300, and 350. These values represented the mean score of the scale and the standard deviation units around the mean. The points were set arbitrarily, to be sure, but they were also convenient in a number of ways. However, whereas the anchor points were statistically useful, they enjoyed no intrinsic value in terms of describing *what should be expected* of students taking the NAEP assessment.

This transition from item difficulty values to the NAEP scales to anchor levels reflects not only advances in the field of measurement and the technology of large-scale assessment, but also changes in the social, political, and educational climate of the United States. Policymakers, legislators, parents, business leaders, and state and local educators, all sensed a need for education standards. The anchor levels met the needs of those who were interested in a normative standard, but the other users of the NAEP data were looking for a criterion-referenced standard, similar to the kind now being reported on the national assessment.

Number and Operations; Measurement; Geometry		Algebra and Functions; Data Analysis, Statistics, and Probability
	0% at or above —300—	(305) Interpret data in bar graph
Word problem, multiply 3 x 1 1/2 (301) Multi-step word problem with money (293)		
Recognizes letter N has parallel lines (288)		
Division word problem with remainder (281)		
Round and add to solve word problem (275) Draw a square, given two vertices (272)		(276) List combinations in simple sample space (275) Explain a pattern 2 x 2, 2 x 2 x 2, etc.
Subtract 503 - 207 = [ ] (267)		(271) Select 5 x [ ] to fit word problem
Select 6 x 24 = [ ] to fit word problem (258)		
Division word problem, $84 by 21 (252)		(256) Circle points on a grid
Round decimal to whole number (248)	17% at or above —250—	
Use ruler to measure length in cm. (241) Two-step word problem with money (238)		(246) Complete table increasing numbers by 50
Subtraction word problem, whole numbers (233)		(236) Read data from a table
Identify 356,097 from its words (228)		
Recognize cylinders (223)		
Subtract 2 numbers with regrouping (215)		
Multiply & divide 2-digit numbers/calculator (205)		
Select instrument to weigh apple (198)	72% at or above —200—	
Divide 108 by 9 / calculator (189)		
Round money to nearest dollar (184)		
Add two 3-digit numbers (178)		(181) Recognize next figure in simple pattern
Multiply 3 x 405 / calculator (156)		
	98% at or above —150—	

Figure 11.1. Percentages of students at or above points on the NAEP 1992
mathematical scale and selected tasks

The National Assessment Governing Board (NAGB), the legislatively-constituted body responsible for setting policy for the National Assessment program (Public Law 100-297, 1988), introduced in 1990 yet another way of describing the performance of students on the NAEP. This method depends largely on the literature in the field of standard-setting, and results in cut scores and descriptions of content at and around the cut scores. These new standards—called *achievement levels*—describe *what students should know and be able to do* on the NAEP assessment at three points in the proficiency distribution. As noted in Chapter 9, these three points correspond to basic, proficient, and advanced levels (NAGB, 1990). Figure 11.2 presents a brief comparison of the salient characteristics of both types of standards—achievement standards and anchor levels.

Characteristics	NAEP Achievement Standards	NAEP Anchor Levels
Nature	Arbitrary points on scale	Arbitrary points on scale
How selected	Reflects professional judgment of representative group	Reflects standard deviation units on a normal distribution
Degree of involvement in setting levels	Representative group recommends the levels; hundreds of professionals and general public participate in planning/designing/implementing level-setting meetings and public hearings; widely disseminated for comment before adoption by NAEP	Professional educators (subject matter specialists) describe levels; widely disseminated when NAEP data are released
Percentages of students at/above levels	Initially varies across content areas; over trend can show growth or decline	Initially fixed across content areas (by definition they delimit the area under a normal curve for all 3 grades combined); over trend can show growth or decline
Interpretation of cut scores on scale	Judgment-bound, i.e., scores reflect student performance in relation to content of assessment at selected points on scale	Empirically-bound, i.e., describes proficiency in relation to national norms, e.g., scores reflect relative standing of subpopulations such as males/females, blacks/whites
Utility	What students *should know and should be able to do*; maximum utility in relating within grade performance to curriculum standards	What students *know and can do*: maximum utility in examining differences among various subpopulations
Relevance	Maximum relevance found in providing conceptual understanding of NAEP scores for policymakers and other users of NAEP data	Relevance limited to describing statistical characteristics of distribution of performance
Validity/reliability	Some evidence of both	Replications of anchoring showed consistency between two half-panels
Evaluation evidence	Internal technical report from contractor studies; external evidence from National Academy of Education evaluation and others	Technical reports available for each assessment cycle which describe anchoring process and internal contractor studies

Figure 11.2. Comparison of NAEP achievement standards & anchor levels

Since the early 1960s the research literature on standard setting in education has grown tremendously in volume. It now covers a wide range of relevant issues related to criterion-referenced testing as opposed to norm-referenced testing. The reliability and validity of test items, the development and selection of assessment options, and the definition of the domains of interest, are a few areas of scholarship.

Although the criterion-referenced movement is generally credited to the seminal work of Glaser (1963), several major contributions to the field have appeared in more recent years. These include the work by Berk (1984), who covered a number of important topics in this area. Important also are an issue of the *Journal of Educational Measurement* (1978) and a special issue of *Educational Measurement: Issues and Practices* (1991), both of which focus on implementation issues. There are also a number of targeted articles that have contributed significantly to the field (e.g., Angoff, 1971; Berk, 1986; de Gruijter & Hambleton, 1984; Forsyth, 1991; Hambleton et al., 1978; Linn, 1980; Shepard, 1980; Shepard, 1984).

Progress has certainly taken place; undeniably much more is known now than in 1963. At the same time, however, there is a certain quality to some of this work. Much of it really does not appear likely to culminate in anything resembling finality or success (Cizek, 1993). For example, a great deal of subjectivity seems to be inevitable in the setting of standards, no matter what approach is taken (Glass, 1978; Linn, 1978). On the other hand, the debates on the issues will be stimulating and relevant to those interested in the setting of content and performance standards.

The debate regarding the utility and technical quality of the new NAEP standards has been prolonged and public. This debate has been informed by a number of evaluation studies that have examined various aspects of standard-setting in the context of a large-scale assessment in education (American College Testing, 1993; US General Accounting Office, 1993; National Academy of Education, 1993). Even though national assessment traditionally has been considered a low-stakes endeavor, the real consequences of setting standards on NAEP are unknown. It is certain, however, that there are threats to internal and external validity. These threats must be considered carefully in planning and implementing a standard-setting activity. Other fundamental considerations moreover apply when such standards are to be used for measuring change. A discussion of some of these issues follows below.

Measuring Trends

The purposes for setting standards in education are multiple. Standards-based reporting of academic achievement supports informed decision-making by local and state school administrators as well as policymakers in the areas of student and program evaluation, teacher accountability, resource allocation, and policy planning. Zwick (1992) identifies five general categories of change that are of interest in measuring trends. Her analysis is developed in the context of NAEP, but could have generalizability in other situations, for example in cases where one is interested in comparing scaled proficiency distributions over time. One of the five categories is of particular interest if the goal is to support valid inferences about academic achievement measured against a standard. This is captured by the question: is the percentage of students at or above a given standard of student achievement greater at point 1 than point 2?

Table 11.1 provides an example of such a change from the 1990 and 1992 NAEP mathematics assessment. One can see significant shifts in the distribution of student performance between 1990 and 1992. The percentage of students not meeting any of the standards (Below Basic category) was significantly reduced at all grade levels. For grades four and eight this resulted in significant increases at the *Basic Level* and the *Proficient Level*; at grade 12, the shift caused a significant change at the lowest standard, *Basic*.

Table 11.1. Percent of students in the U.S. at or above the NAEP mathematics standards

Grades	Assessment years	Basic	Proficient	Advanced	Below basic
4	1992	61(1.0)>	18(1.0)>	2(0.3)	39(1.0)<
	1990	54(1.4)	13(1.1)	1(0.4)	46(1.4)
8	1992	63(1.1)>	25(1.0)>	4(0.4)	37(1.1)<
	1990	58(1.4)	20(1.1)	2(0.4)	42(1.4)
12	1992	64(1.2)>	16(0.9)	2(0.3)	36(1.2)<
	1990	59(1.5)	13(1.0)	2(0.3)	41(1.5)

How Many Standards Should There Be?

The issue of whether to have a single standard or multiple standards is hardly ever addressed, except by assumption. For example, few school districts use a Pass/Fail decision to evaluate student performance in course work like algebra or geometry. The more typical grading schemes employ an *A-to-F* decision, or a numerical equivalent of this, *95-to-60* or less. It is assumed that multiple points are necessary to adequately describe student performance in the course. Likewise, the new move to develop "alternative" assessments has generated scoring protocols that generally employ a polytomous grading scheme representing the distribution of performance in the content areas. Again, it is assumed that there is some benefit to be gained from scoring writing samples not as *right/wrong* but onto an appropriate *multiple-point* scale, for example a four-category scale with the levels undeveloped, minimally developed, fully developed, and elaborated.

It has been determined by the National Assessment Governing Board, which is responsible for NAEP reporting, that the process of setting achievement standards on the NAEP should produce three achievement levels for each content area and at each grade level assessed. The desire was to describe the full range of student performance, including the students whose performance is in the mid-range, as well as both those whose performance is below and above that range. This desire was based on the notion that it is highly important to hold realistic expectations of the level of performance the students can achieve, no matter what their achievement level might be at the time the assessment is administered. The Board concluded on the basis of research studies that the three benchmarks set on the NAEP scales would offer sufficient scope for all students to meet realizable expectations. Another concern was the commitment to preserve trend results in the NAEP. It was deemed that the three achievement levels would accommodate growth (and possible decline) in all the ranges of the performance distribution.

Of course there are some requirements that need to be accommodated. First, the range of item difficulties on the assessment must be such that it spans the lower and higher ends of the distribution, as well as provides solid coverage in the middle of the performance range. Ideally, if one were to employ the results of the standards-based assessment to make high-stakes decisions about the individual examinees, then the items to be included in the test should discriminate well at the cut scores. For reporting distributional characteristics, however, this requirement is relaxed somewhat, but not entirely. Second, there should be clear distinctions between multiple standards for any given cohort. So, for example, if there are two or three standards within a single grade, then these should be substantively different from each other. On the NAEP scales, the differences between adjacent levels in any one grade is about one-half standard deviation, and this seems to represent substantive differences between the levels.

How Frequently to Measure Change?

How frequently one should measure change is a function of a number of conditions affecting the assessment. First, consider the cognitive demands of the standards. On the one hand, if the assessment measures mastery of fairly simple skills and behaviors—such as those generally acquired during the early childhood years—then measures of a trend could conceivably occur frequently. If, on the other hand, the cognitive demands of the standards are more complex, such as higher order thinking skills generally acquired in a later developmental period, then this would dictate less frequent measures of change, since success in meeting the standard would normally require more time for complex skill acquisition.

In addition to cognitive demands, frequency is also a function of the level of focus of the standards—individual assessment or group monitoring. If the purpose is to estimate growth for an individual then more frequent estimates are desirable. Presumably, the kinds of decisions informed by growth curves for individuals are different from those informed by trend lines for groups. Diagnosis, placement, and formative evaluation are a few of the decisions appropriate for individuals which require more frequent measures of change. In contrast, accountability, curriculum evaluation, and policy planning decisions can be equally informed by long-term, less frequent measures of change against a standard.

Stability and Change in Measuring Trends

In general, the larger the number of measurement points in time the better change can be assessed. Two measurement points, at t_1 and t_2, may indicate the beginning of a trend, but it may also indicate an anomaly as in the case of the 1986 NAEP reading assessment (Beaton & Zwick, 1990). It is highly desirable when assessing change to ensure that there is stability across time: in the measuring instrument, the administration procedures, and the types of data analyses that are conducted. However, a tension between stability and continuity with the past has developed over the years, and calls for state-of-the-art assessments that allow the NAEP to "push the envelope" of assessment technology and design continue to be heard. The NAEP maintains stability by employing various samples in its design. The main NAEP samples are used for current assessments. These cross-sectional samples offer the basis for current estimates of student proficiency and of short-term trends. The estimates of the long-term trends in student performance are based on linked and comparable samples of examinees, and the data collection relies on the same instrumentation and administration procedures as those used in the first year a particular assessment was conducted—which, in some cases, dates back to 1969. The NAEP also employs special purpose samples, some of which are specifically designed to accommodate special research initiatives or necessary and unavoidable changes in an assessment. For example, the hand calculator used in the 1990 mathematics assessment was no longer being manufactured two years later. Therefore, a similar though not identical hand calculator had to be employed in the 1992 assessment. Stability was ensured by employing a bridge sample which estimated the effects of such a change, and allowed linkage to the main assessment. Table 11.2 shows the various types of samples used in the 1992 NAEP mathematics assessment (Johnson et al., 1992).

Similar sample types are also employed in the assessments of reading and writing. In 1992, 154 distinct samples with a quarter of a million examinees were employed to effectively measure student achievement in the United States. This produced, first, cross-sectional measures of student proficiency in mathematics, reading, and writing in three age by grade cohorts, second, short-term trend data for mathematics achievement, and third, long-term trend data in reading, writing, mathematics and science achievement spanning the first 23 year

Accommodating changes

Accommodating changes either in the content or in the performance standards requires vigilance over time. If curriculum content changes radically, then obviously the assessment must be brought into line with those changes. What constitutes radical change is, however, a matter of judgment. For example, the National Council for Teachers of Mathematics (NCTM) proposed in 1989 what most professionals in the field have judged to be substantial changes in the scope and sequence of the mathematics curriculum and the way it is taught (NCTM, 1989). Concurrently with the NCTM initiative a new NAEP mathematics assessment framework was being developed. Since these efforts were occurring simultaneously, the NCTM content standards were not fully reflected in the 1990 NAEP framework. This was an unfortunate lack of synergy that subsequently had to be addressed through a series of revisions in the NAEP assessment framework. However, the revisions needed to be constrained to some degree because the 1990 framework represented the beginning of a new 'trend line' in the NAEP, and those changes which would have caused a change in scale were not considered.

Table 11.2. NAEP 1992 examinee samples in mathematics

Sample	No. text booklets	No. item blocks	Mode	Cohort assessed	Sample size
01[Math-Main]	26	13	Print	Age 9/grade 4	10,183
02[Math-Main]	26	13	Print	Age 13/grade 8	10,183
03[Math-Main]	26	13	Print	Age 17/grade 12	10,183
01[Math-Main Est]	1	4	Tape	Age 9/grade 4	2,350
02[Math-Main Est]	1	4	Tape	Age 13/grade 8	2,350
03[Math-Main Est]	1	4	Tape	Age 17/grade 12	2,350
01[Math-CalcBr]	1	3	Print	Age 9/grade 4	2,350
01[Math-State]	26	13	Print	Grade 4	112,500
02[Math-State]	26	13	Print	Grade 8	112,500
01[Math/SciBridge86]	3	9	Tape	Age 9	6,000
02[Math/SciBridge86]	3	9	Tape	Age 13	6,000
03[Math/SciBridge86]	2	6	Tape	Age 17	4,000

Source: Technical Report of the NAEP 1992 Trial State Assessment Program in Mathematics.

The NAEP learned another difficult lesson from the 1986 reading anomaly, and now subscribes to the axiom, "When measuring change, do not change the measure," (Beaton & Zwick, 1990). Change in scale is an empirical question that can be answered by examining the stability of the linkages between t_1 and t_2. In the case of the 1990 mathematics framework, new exercise formats were added to the 1992 and 1996 item pools in the form of extended constructed response items. Empirically these effectively raised the ceiling on the assessment so that some items would scale at higher levels in the proficiency distribution, thus accommodating the *Advanced Level* standard. However, one should not oversimplify the case for stability in measuring change. Even maintaining identical exercises in an assessment from year to year does not necessarily guarantee growth measures free from confounding influences. For example, a particular item may function in a certain way at t_1 in terms of the kinds of examinee skills and behaviors elicited; and at t_2, the processes used by the examinee to solve the identical problem may be quite different.

Use of linking items

Using a common set of linking items in estimating trend is a well-known procedure in the NAEP. As mentioned earlier, administering the trend assessment using long-term trend samples provides stability across time; linking items ensure comparability across varying forms of the assessment. Typically, in the NAEP no examinee is administered the entire exercise pool. Large numbers of items are employed to ensure adequate content coverage in the domain. Both the exercises and the examinees are matrix sampled. Items are clustered into "blocks" (units of about 10 to 15 items) which are randomly assigned to examinees. About 60 percent of the assessment blocks carry forward from one cycle to the next. For example, from the 1990 mathematics assessment, four of the seven blocks carried forward to 1992 as trend blocks. In 1992, nine blocks (out of the 13 blocks in the 1992 main assessment) will carry forward to the 1996 assessment, Y from the 1990 assessment, and Z from the 1992 assessment. Items are packaged in clusters called 'blocks' and are cycled in and out of the assessments as blocks to minimize order effects. In subjects such as reading, they must be treated as clusters because generally a series of items accompanies a single reading passage. The intent is to retain a sufficient number of items from year to year so that the scale can be equated and the trend preserved. In the NAEP, depending upon the size of the item pool, the requirement could vary from as few as 60 items to as many as 125 items. If the analyses require linking for several subscales, as in the case of the mathematics assessment, then the number of items required will be larger than if a single scale or only a few subscales are reported.

The Case For or Against Adjusted Scores

There are both policy and technical arguments in this area of measuring change in standards. Mosteller and Tukey (1977) offer a discussion of two approaches, *direct and indirect standardization,* which can be used to adjust scores in a population in a fashion similar to the handicapping principle used in golf and other sports. Both of these approaches have been examined in the NAEP context as ways of 'leveling the playing field' for the 40 or so jurisdictions participating in the Trial State Assessment (TSA) component of NAEP (Mazzeo & Johnson, 1994). States vary in average proficiency due to a variety of factors, some of which are not in their own control. Direct standardization answers the question, what would the states' proficiency distributions look like if the population characteristics of the states were similar to those of the nation as a whole? In this case, the nation would become the common reference population to which each state would be compared. Consequently, each state would have both raw scores and adjusted scores.

In the *indirect* approach a slightly different question is asked. What would be the nation's proficiency if the nation had population characteristics similar to that of each state? In this case, the nation would have as many 'adjusted values' as there are jurisdictions being compared. Table 11.3 displays the comparative results of indirect standardization using a *weighting method* with a *regression approach*.

Table 11.3. 1992 NAEP trial state assessment adjusted scores in reading

Means	Jurisdictions				
	A	B	C	D	E
Observed state mean	189	200	209	219	229
Observed national mean	216	216	216	216	216
Weighted national mean	196	207	213	219	223
Regressed national mean	193	204	211	219	225

In the *weighting method,* Mazzeo and Johnson (1994) defined a 4x4x2 matrix by cross-tabulating four levels of two variables, namely race/ethnicity (RET) and type of community (TOC), with two levels of a third dichotomous variable, limited English proficiency (LEP). In the *regression method* a substantially larger number of variables can be successfully entered into the adjustment equation. For the purposes of the analysis conducted by Mazzeo and Johnson (1994), the nine variables included, in addition to RET, TOC, and LEP, gender, parents education, amount of reading materials (books, magazines, newspapers) found in the home, percent school minority enrollment, presence of an individualized education plan (IEP), and the percent of federally subsidized lunch in the school.

The weighting method limits the number of variables accounted for in the adjustment because the cell sizes become very small quickly. On the other hand, the regression method allows a greater number of variable to be used in the adjustment, even though order of entry is a critical decision.

There are policy arguments on both sides of this issue. Adjusting scores is an attempt to simplify and clarify what would otherwise be complex data. Adjustments are aimed at improving the utility of scores for particular audiences who are not technically sophisticated and find adjusted scores more intuitively appealing. There is also something to be said for providing data for normative comparisons that are fair and even-handed.

On the other hand, the nuances of adjusted scores may elude the casual user, and misinterpretation may result. Moreover, adjusted scores may serve no purpose in an age of world-class standards since all of these methods could have the effect of lowering expectations for some students. In Table 11.3 jurisdiction A has a very large percentage of minority students. The adjusted national score says, if the nation had as many minority students as jurisdiction A then it, too, would be achieving less well—down 20 to 23 score points. Jurisdiction E is a fairly homogeneous and mostly rural population. The adjusted national score says, if the nation were homogeneous in composition with little distinction between urban, suburban, and rural subpopulations, then it, too, would be achieving better—up 7 to 9 score points.

Part of the intent in setting standards is to reach consensus on what is important to be taught and learned (*content standards*), to decide how much of that must be achieved by students (*performance standards*), and to measure it in the best way possible, reporting on how well students achieve against the standards. If a different yardstick which depends upon the demographic characteristics of the population group or family structure is used to measure achievement, then what kind of solid inferences can be made about students reaching the standards?

Conclusion

Setting standards for educational achievement is a treacherous business; measuring growth against such standards is even more treacherous. This chapter has identified some of the internal and external threats to the validity of measuring change. Issues such as assessment design, data scaling, and the decision consequences of standards, not only may be threats to validity, but also impact on the inferences one can make from national assessment data. The chapter has also discussed some of the critical questions that must be examined in the context of measuring change against the standards, including the frequency of measuring change, stability and change when measuring a trend, and the use of linking items in measuring change. Finally, the chapter has explored adjusted scores for making comparisons among jurisdiction having different population characteristics.

References

American College Testing. (1993). *Setting achievement levels on the 1992 National Assessment of Educational Progress in mathematics, reading, and writing: A technical report on reliability and validity*. Iowa City, IA: American College Testing Board.

Angoff, W.H. (1971). Scales, norms, and equivalent scores. In R.L. Thorndike (Ed.), *Educational measurement*. Washington, DC: American Council on Education.

Beaton, A.E., & Zwick, R. (1990). *The effect of changes in the national assessment: Disentangling the NAEP 1985-86 reading anomaly* (Report No. 17-TR-21). Princeton, NJ: Educational Testing Service.

Berk, R. (1984). *A guide to criterion-referenced test construction*. Baltimore, MD: Johns Hopkins University Press.

Berk, R. (1986). A consumers guide to setting performance standards on criterion-referenced tests. *Review of Educational Research 56*, 137-172.

Cizek, G. (1993). Reconsidering standards and criteria. *Journal of Educational Measurement 30*, 93-106.

de Gruijter, D.N.M., & Hambleton, R.K. (1984). On problems encountered using decision theory to set cutoff scores. *Applied Psychological Measurement 8*, 1-8.

Educational Measurement: Issues and Practices (1991). Vol. 10(2).

Forsyth, R.A. (1991). Do NAEP scales yield valid criterion-referenced interpretations? *Educational Measurement: Issues and Practice 10,* 3-16.

Glass, G. (1978). Standards and criteria. *Journal of Educational Measurement 15,* 237-262.

Glaser, R. (1963). Instructional technology and the measurement of learning outcomes. *American Psychologist 18,* 519-521.

Hambleton, R.K., Swaminathan, H., Algina, J., & Coulson, D.B. (1978). Criterion-referenced testing and measurement: A review of technical issues and developments. *Review of Educational Research 48,* 1-47.

Johnson, E.G., Mazzeo, J., & Kline, D.L. (1992). *Technical report of the NAEP 1992 trial state assessment program in mathematics.* Washington, DC: National Center for Education Statistics, US Department of Education.

Journal of Educational Measurement (1978). Vol. 15(4).

Linn, R.L. (1978). Demands, cautions, and suggestions for setting standards. *Journal of Educational Measurement 15,* 301-308.

Linn, R.L. (1980). Issues of validity for criterion-referenced measures. *Applied Psychological Measurement 4,* 547-562.

Mazzeo, J., & Johnson, E. (January, 1994). Standardization and regression-based estimates of state comparison values. Unpublished paper presented at a meeting of the Design and Analysis Committee, Educational Testing Service, Washington, DC.

Messick, S., Beaton, A.E., & Lord, F. (1983). *A new design for a new era.* Princeton, NJ: Educational Testing Service.

Mullis, I.V.S., Dossey, J.A., Owen, E.H., & Phillips, G.W. (1993). *The NAEP 1992 mathematics report card for the nation and the states.* Washington, DC: National Center for Education Statistics, US Department of Education.

Mosteller, F., & Tukey, J.W. (1977). *Data analysis and regression: A second course in statistics.* Reading, MA: Addison-Wesley Publishers.

Phillips, G.W., Mullis, I.V.S., Bourque, M.L., Williams, P.L., et al. (1993). *Interpreting NAEP scales.* Washington, DC: National Center for Education Statistics, US Department of Education.

Public Law 100-297. (1988). National assessment of educational progress improvement act (Article No. USC 1221). Washington, DC.

Shepard, L. (1980). Standard setting issues and methods. *Applied Psychological Measurement 4,* 447-468.

Shepard, L. (1984). Setting performance standards. In R. A. Berk (Ed.), *A guide to criterion-referenced test construction.* Baltimore, MD: Johns Hopkins University Press.

US, General Accounting Office (1993). *Educational achievement standards: NAGB's approach yields misleading interpretations* (GAO/PEMD-93-12). Washington, DC: Author.

US, National Academy of Education (1993). *Setting performance standards for student achievement.* Stanford, CA: Author.

US, National Assessment Governing Board (1990). *Setting appropriate achievement levels for the National Assessment of Educational Progress: Policy framework and technical procedures.* Washington, DC: Author.

US, National Commission on Excellence in Education (1983). *A nation at risk: The imperative for educational reform.* Washington, DC: US Department of Education.

US, National Council for Teachers of Mathematics (1989). *Curriculum and evaluation standards for school mathematics.* Reston, VA: Author.

US, National Governors Association (1991). *Educating America: state strategies for achieving the national education goals.* Washington, DC: Author.

Zwick, R. (1992). Statistical and psychometric issues in the measurement of educational achievement trends: Examples from the National Assessment of Educational Progress. *Journal of Educational Statistics 17,* 205-218.

Chapter 12

International Standards for Educational Comparisons

ANDREAS SCHLEICHER

IEA Secretariat, The Hague, The Netherlands

This chapter discusses the type and nature of mainly technical standards that need to be applied in data collection activities intended to yield valid and reliable information for use in international comparisons of education systems. Attention is given both to standards that help ensure and verify the adequacy and quality of results of data collections, and to standards concerned with the data collection operations. It is concluded that these criteria should be consistent with the selected comparative approach, the type of variables to be measured, the nature of the data to be collected, and the populations for which the data are to be analyzed and reported.

There is an apparent and growing demand for valid, reliable and comparative information at the international level that can offer insights into the development, functioning and performance of education systems, and that can assist in the planning and management of educational services. Given the variation in the education systems of different countries, it is essential that international comparisons are based upon a common framework of standards which are agreed upon prior to the data collection, and against which the adequacy of the data collection procedures and the accuracy of the results can be validated. The purpose of this chapter is to show why such international standards are important, for what areas standards are needed, how they may be derived, and how they can be used to secure a general and internationally valid understanding of the aspects of education that are measured and, second, to ensure that the data collection activity yields information that can answer the intended policy and research questions and that the data collection activity is undertaken cost-effectively and under comparable conditions.

Importance of Standards in International Comparisons

Technological developments have affected job markets throughout the world since the 1970s. Consequently, education policymakers increasingly see the need to anticipate the demands of a future labor market and to plan the supply of educational services accordingly. The policymakers who are faced with this challenge require information. Questions they typically ask are: "How good or bad is the performance of our education system?"; "Where are the weak points?"; "What should be done to remedy the weaknesses?"; "What will it cost?"; and "How long will it take?". For practical and ethical reasons, large-scale experimental studies to answer these questions cannot be conducted. Hence researchers and policymakers have come to rely on the variation that exists both within and between the education systems of the different countries. For reasons mentioned in Chapters 1 and 2, the number of countries seeking to take part in international and comparative studies is increasing. This shows a growing demand for valid, reliable and comparative information that can be used to examine change over time and monitor educational progress.

Several prerequisites for obtaining the required comparative information apply. The first is that the populations and the variables to be measured are in fact comparable within a well-defined conceptual framework. Another condition is that the employed methods and procedures should allow for fair comparisons. A third requirement is that the information must be obtained in a suitable and cost-effective way. This is especially important because both the data producers in the countries and those coordinating such international activities must devote substantial resources to the establishment of communication channels to data providers of different locations and organizations, to the data collection and to the analysis and dissemination of the data (OECD, 1993).

It is not easy to obtain comparable information on education. This is complicated by the fact that the educational policy priorities and the structures of the governance of education differ between countries. Education systems cannot, moreover, be held to a fixed position and it is difficult to ensure consistency in the demand and supply of information because education systems change over time. Given the divergences in goals and interests, and the lack of a generally accepted representation of relationships in the education sector, international comparisons require that a common framework of standards is agreed upon prior to the collection of data, Standards in this context are understood both as the *principles* to which an international data collection operation should conform, and as the *measures* by which the quality and accuracy of the results can be judged. Such standards can ensure that the conceptual framework underlying the research and policy questions is applicable in all participating countries; the populations and entities of reporting which are being compared are in fact comparable; the definitions underlying the variables to be measured and their operationalization in the data collection instruments are valid and equivalent in the different education systems; the actual measurement of the variables is consistent across the involved countries; the criteria for judging the accuracy and precision of the results are comparable; the data collection and data management operations are undertaken under comparable conditions; and that the data analyses undertaken are in accordance with both the type and quality of the collected data and the requirements of reporting. Finally, standards in international comparisons can assist in the interpretation of the results.

All research dealing with human populations is subject to a wide range of economic, socio-cultural, and administrative and practical constraints. Some of these constraints, for example those which can be expressed in economic terms, can be addressed in a way that is amenable to algorithmic expression. Others, for example "ethnicity" require a more subtle form of measurement and analysis. The final design of a data collection activity usually represents a compromise position that seeks to satisfy most of these external constraints. The definition of standards can ensure that the constraints have a similar impact on the results of a data collection activity in the different education systems and that the available resources are used so as to provide optimal results within the given constraints.

Nature and Type of International Standards

The nature and type of standards that are required for international comparisons of education systems depend on the following criteria: the purpose, approach and nature of the comparisons that are undertaken; the variables that are measured; the data that are used to measure these variables; the entities for which the data are collected and reported; and the design of the data collection operations.

Standards can be set at various points in time and by various authorities in a data collection activity. Commonly, they are defined by the *data requesters*—that is, by the authorities or international agencies that formulate the initial questions and that initiate the data collection. These groups are often also responsible for the reporting and the dissemination of the results obtained from the data analysis. The standards must be specified so that they fulfill the requirements of the *data users,* that is, the national authorities, researchers, media, and other members of the public who will eventually make use of the results. The standards are applied and verified by the *data producers*—usually the authorities that conduct the data collection and are responsible for the management, processing, and analysis of the information. Finally, the standards must be implemented by the *data providers*—the organizations, authorities, or individuals who supply the actual data to the international or national agencies that initiated the activity.

It is useful to distinguish between two types of international standards. The first refers to the standards that ensure the quality of the data, the legitimacy of the comparisons, and the adequacy of the results and interpretations emerging from the comparisons. The standards in the second category ensure that the processes that are followed in order to develop good measures and to obtain useful, accurate, and comparable information are adequate and can be regarded as equivalent for the countries involved.

International standards for data quality and results depend on the aims and the design of a data collection activity, as well as on the various economic, socio-cultural, and administrative constraints under which the data collection activity is conducted. General standards can therefore not be established without information on the—often competing—constraints as well as information on the intended use of the results. Indeed, a certain standard that may be adequate for the comparison of certain statistics at a particular level of reporting may, at the same time, be too low as to show significant differences when comparing other statistics or when another level of reporting is used. Or the standard may be unnecessary high with respect to the intended comparisons and thus be a waste of project resources. For example, the requirement that survey samples in international comparative studies should result in an effective sample size of, say 400, for mean student achievement scores may be too low as to show significant differences between the national mean scores of similar education systems. It may also prevent meaningful comparisons at a lower level of aggregation or meaningful comparisons of the results for certain subgroups. In another situation, an effective sample size of 400 may be unnecessary high if the main aim is to establish a rank-order of countries from very different levels of student achievement; or it may result in a situation where the sampling errors are very small compared with non-sampling errors, in which case project resources would be better invested in improving the quality of the measurement and reducing the non-sampling errors rather than aiming for small standard errors of sampling (Ross, 1988).

The need for standards for results also depends on the type and amount of data that are reported. To take the previous example further, if national mean ability estimates based on survey samples are reported together with their standard errors, then the user of the data can evaluate the precision of the estimates, and take this information into account when interpreting the results. If no standard errors are reported, however, then it is essential that the data meet specified minimum requirements of precision, so that misinterpretation can be avoided.

Often the project results and their implied metric cannot be interpreted without a detailed knowledge of the measurement process and the underlying variables. For example, a mean score of 500 for a sample of 14-year-old students in a country on a particular test does not provide immediate information about the actual level of achievement in that country, unless the test items and the conditions under which they were administered are known to the user of that information. In such cases, it can be helpful to assign external reference points to the implied scales, so that the levels of achievement can be anchored with respect to external criteria. It can also be useful to define performance standards for the observed achievement scores. Due to a variety of factors—for example, differences in culture, demands, resources, and wealth—the desired performance standards may not be the same in all the countries participating in an international study (Elley, 1994). The proportion of the students who perform below a certain performance standard may also vary across countries, and countries that have high rankings in terms of mean student performance are not necessarily those that have the lowest percentage of students below their own benchmarks. The establishment of performance standards can therefore contribute to the evaluation of national outcomes and the interpretation of national differences.

Standards for processes need to account for differences in the underlying definitions of the measures as well as for different procedures used for the data gathering and processing in the participating countries. They need to cover the four main stages of the data collection activity: design, data collection operations, data preparation and processing, and data analysis and reporting.

Standards for processes can be established as general principles for a data collection activity, and can be established to some extent independently from the particular aims and design of the data collection activity. Such standards are, however, often highly interrelated, and standards in earlier phases of a data collection activity often have an impact also in later phases. Standards for processes have often also an impact on the use of the survey results. For example, the kind of sampling operations that are undertaken to obtain data will constrain the type of analysis that can be undertaken with the data and therefore the results that can be obtained. If, for instance, the design of a sample is optimized in order to provide optimal national estimates, then this design may not be adequate to obtaining suitable results on a lower level of aggregation, such as those required for comparisons between provinces or regions.

Standards and the Units of Data Collection

Standards and Target Populations

Each data element is collected with an explicit or implicit reference to an underlying entity (such as an education system, a school, a classroom, or a student), which is referred to as the *base entity*. For example, achievement scores can be associated with students, enrollment rates can be associated with schools, etc. International comparisons must rely on standards which ensure the comparability of the entities for which data are collected and data are reported. The entire set of these base entities comprises the target population. Since it may not be possible to obtain data from all the entities for which data are ideally sought, a distinction is therefore made between three classes of target populations: the *desired* target population as the population for which results are ideally required; the *defined* target population as the population which is actually studied and whose elements have a known and non-zero chance of being included in the data collection; and the *achieved* target population as the population for which data are actually obtained (Ross, 1987).

Each data collection activity must commence with a precise definition of the target population in terms of four criteria (Kish, 1965):

1. The *content* of the population—for example, "predominantly publicly financed private schools" where, of course, terms like "predominantly public", "financed" and "private" need to be operationalized;
2. The *operationalized base entities* for which data are collected;
3. The *extent* of the population—for example, "all schools of all educational subsystems of a country with students enrolled on a full-time basis"; and
4. The *time* to which the definition refers—for example, certain expenditures may refer to a certain financial year or the age of students may refer to a particular day in the year.

The definition of the target population depends, of course, on the type of policy and research questions that are asked. Hence the standards for the population definition are related to the aims and design of the data collection activity. In some cases, for example, a data collection may ask for enrollment rates by grade whereas in others, enrollment rates may be required classified by age. The operational definition of the extent of populations can be very complex depending on the required comparisons and the hierarchical and spatial structure of the populations which are being investigated. In order to obtain unbiased estimates it is necessary to establish standards which ensure that the existence and identity of every entity in the defined target population is known and that the measures required can be obtained from every entity in the population.

If the standards for the definition of the target populations are imprecise or ambiguous, these ambiguities will have an impact in many phases of the data collection and can lead to substantial problems in the comparability of the desired target populations, and thus of the overall survey results (Tuijnman, 1994). The fact that an imprecise population definition may cover a wide range of differences in population definitions across countries often means that the results of a data collection activity refer in fact to populations that are different in kind. For example, if the content of a population definition refers to "mainstream schools" without a further qualification, then this may cover schools or classrooms with unusual curricula, or students with certain disabilities in some countries but not in others, and it may be interpreted differently by different jurisdictions or ethnic groups.

A further potential problem in international data collections is that the target populations that are defined in the different countries may not entirely reflect the internationally defined target population. Countries often need to make modifications and exclude certain parts of the defined target population in order to make the implementation feasible in the context of their national school systems. A data collection may, for example allow the exclusion of schools which are geographically or politically inaccessible or at least very costly to include, allow the exclusion of certain types of schools or regions in a country that are considered to be operating under unusual curricular arrangements or unusual media of instruction, and/or teaching methods, or allow the exclusion of certain handicapped students from the administration of a test.

In order to estimate the impact on the resultant data and to ensure comparability, precise standards must be provided that operationalize the conditions and regulate the exclusion of population elements in the countries. For example, the understanding of "handicapped" students may include different kinds of physical, emotional, and mental disabilities in different countries and, therefore, may vary considerably between countries. There should also be standards which regulate the implementation of population exclusions, for example, it may be required that the exclusion of students within the schools can only be authorized by the school principal or by other qualified staff members and not by the class teacher or test administrator. It is further important that there are standards which regulate the overall extent of population exclusions so that at least an international upper boundary for the potential bias resulting from population exclusions can be established. For example, if the excluded population exceeds 10 percent, then this can have an unacceptably large impact on the accuracy of the results—if these ten percent of the population differ from the rest of the population in a systematic way.

Standards and Units of Selection

A precise definition of the entities underlying the target population is a prerequisite for the definition of the target population. For example, if data are collected for schools or students, then there must be an indication of the criteria that constitute a corresponding school or student in the respective context. Standards must therefore be established that define the characteristics and criteria of the entities that are to be part of the target population. For example, when data about the number of students enrolled on a full-time basis are collected, then internationally valid criteria for the operationalization of the term "full-time" need to be established.

Often the base entities are only implied and not explicitly specified. For example, a question may ask for the enrollment rates of students in a particular school system without defining the service providers that are to be included. Without such specification, different countries may include or exclude different types of schools which thus introduces a potential for bias into the results. Sometimes the lack of specification and standards results in confusion as to what base entities are implied. To take the previous example further, if questions refer to enrollments without operationalizing the entities that are to be counted, then one data producer may count students as individuals whereas another may count enrollments as such. This can result in discrepancies, because in some education systems students may at the same time be enrolled in different educational programs.

A particular problem arises when there are no standards that ensure that the terms used to characterize the base entities share a common meaning across education systems. For example, the distinction between *"private* schools" and *"public* schools" may be well-defined within a certain education system so that it is possible to classify each school in the target population in that country unambiguously as either "private" or "public". But it could be that these terms have a different meaning in different countries and thus have no international validity. For example in one country, the distinction between public and private schools may involve the political or administrative control over decision-making; in another to the extent in which funds are derived from public or private sources; and in yet another to the degree of institutional dependence on funding from government sources.

Standards and Units of Reporting

The entities that are used to define the target population are generally also the entities for which data are collected. However, these entities may not necessarily be identical to those that are to be employed in the subsequent data analysis and reporting. For example, a sample of students may be selected, but for the data analyses the student estimates may be aggregated to form class mean or school mean estimates, or to national aggregates. If the base entities are not identical to the entities of reporting but they are aggregates of these then great care must be taken to ensure that the design specifications, especially the precision requirements, are based on the entities of reporting and not on the base entities. Sometimes results are required for indicators involving multiple base entities, or indicators may require entities of reporting from different levels of aggregation. In this case clear standards are required with respect to the methods used to calculate these indicators and the interpretation of the results.

Particular problems can arise when data for complex entities of reporting—such as data for education systems or sub-systems—are not derived from their base entities but are collected directly from highly aggregated sources. For example, enrollment data may be taken from central registers rather than aggregating them from the respective schools. In such cases, standards need to ensure that in the figures available only those entities are included that are in fact part of the respective target population. Macro-level data also bear potential problems with respect to incompatibilities of the underlying base entities. Often, for example, data requesters ask for break-downs of the data according to predefined classification schemes, even though in many cases it is not possible for some countries to apply these classification schemes with existing data sources. Thus there is a danger that differences in the classification schemes are not accounted for in the presentation of the results.

Standards and Criterion Variables

The aim of the data collection is to obtain information about the unknown population parameters that are under investigation and that are operationalized as functions of these criterion variables. Standards are required for the definition of the criterion variables in terms of their content, classification categories, units of measurement, and a description of their intended use.

The required standards depend on the type of statistics that need to be calculated. These may range from simple descriptive statistics to complex structural relationships. A description of the use of the survey variables should include the required statistics and the intended comparisons. In ideal terms an outline of the analysis plan would be available at the time of the preparation of the design, and the criterion variables would be defined within the context of this analysis plan. It is also important that standards for the operationalization of the variables are specified. This can be done by specifying the methods of observation in terms of both data collection and data analysis. For example, a variable such as "student age" may be directly observable and measurable whereas a variable like "student achievement in mathematics" may only indirectly be measured. Furthermore, information is required concerning the scale type of the variables.

A priori Versus Post-hoc Standards

Where data collections can be specially designed for answering particular policy and research questions, standards can be defined *a priori* in such a way that the data collection methodology as well as the design of the procedures and instruments provide results that are adequate for answering the intended questions.

In other cases, data are collected not in specially designed and centrally coordinated surveys but as part of different national data collections; or existing data are used that initially were collected with different aims and a different design of the data collection. In these cases the adequacy, equivalence, and quality of the respective data needs to be examined *post-hoc*. In such cases it must be demonstrated that the desired data are of sufficient comparability, applicability, and usefulness. Sometimes, post-hoc standards are also required for the interpretation of the results of data collections, for example, when it needs to be ascertained whether a given difference between the countries can be regarded as significant with respect to a particular criterion.

Standards for Data Quality and Results

Standards for Accuracy and Precision

The planning of any sample survey is always dominated by the accuracy that is required for the estimates. The term *accuracy* refers to the size of the deviations of an estimate that is obtained through the data collection from the true population parameter. Decisions in this area are inevitably linked to the purposes for which the results are to be used and the resources available. Sometimes the standards for the required accuracy can be established by a clearly specified statistical decision. In most cases, however, the data collection serves multiple purposes and the different accuracy requirements have to be carefully balanced among these. For example, if in a sample survey a country with a decentralized school system intends to compare different regions or provinces, then it would be best to draw equal sized samples from the different regions or provinces in order to minimize the standard error of the difference between the subgroup means. However, if only national estimates are of interest, then a design in which the sample sizes proportionally represent the size of the regions or provinces is usually more efficient. If a sample survey must serve both purposes, then a compromise between both designs has to be sought.

Sometimes, and especially in cases where data collections are based on probability samples, standards are defined which separately take into account the error that is attributable to the selection of the sample but which does not account for other sources of errors, such as measurement errors, or errors from non-response. In such cases, standards are established for the *precision* rather than the accuracy of the results; where the term precision refers to the size of the deviations from the estimate which would be obtained through an infinite replication of the sample. Probability samples allow inferences from sample estimates to population values and provide estimates of the precision, based on measures of variability that can be computed from the sample data for the statistic for which estimates are sought.

Precision standards can be formulated using various approaches. *Confidence intervals* specify an interval into which the desired sample estimate is required to fall with a specified probability. The disadvantage of this approach to defining precision standards is that it depends on the statistical estimator and the distribution of the criterion variable, which then additionally needs to be known. Precision standards can also be defined as the *maximum tolerable standard error* of the desired sample estimator relative to the population variance of the criterion variable. Third, precision standards may be defined through *effective sample sizes*. This approach keeps the distributional characteristics of the criterion variable separate from the procedure used to set standards.

It is possible that data are provided only for highly aggregated entities, such as schools of a certain type in a certain education system, or schools at a certain level of education. A problem can arise if there is no information on the precision of such data because, since if there is no variability, such indications cannot be derived from the data and therefore must be established from external sources. If the data values are estimates then serious misinterpretation can occur if information on the estimation procedures, the quality of the estimation, and the tolerance limits is not taken into account. If for the variable under investigation there is only little variation between the countries, then small inaccuracies in the data for individual countries can result in rank-order shifts between the countries.

In order to ensure the accuracy of the data, standards for determining measurement errors and errors resulting from nonresponse or non-coverage are also required. Whereas nonresponse usually presents a problem only in sample surveys, noncoverage can also be a problem in macro-level data collections. For example, data on private schools may not always include data for a small proportion of independent private schools.

Standards for Reliability

Each measurement is derived from a particular set of questions or observations and is obtained under a particular set of circumstances. Thus the extent to which the observations obtained are affected by measurement error—the error associated with all those influences which are not intended to be measured—must be determined. Clearly, in order to determine to what extent the results can be generalized with respect to a broader underlying universe of questions and observations, standards need to be established for the *reliability* of the measures. Different types of reliability standards were discussed in Chapters 6 and 10. Their application depends on the type of errors that need to be accounted for (Livingston, 1988). The design of a data collection activity must ensure that a systematic analysis of variance can be undertaken as a prerequisite for a complete understanding of the error sources.

Where results are derived from multiple variables, an estimate of the reliability can often be obtained as a measure of the internal consistency of the variables (for example, the correlation between the results obtained from two halves of the same set of variables or the average of the correlations of the individual variables with the composite measure). A key problem is the estimation of the reliability for macro-level data, such as data on educational finance, where the variance is unknown and therefore the internal consistency cannot be examined.

Numerical standards of reliability depend on the aims of the data collection activity. Also here standards need to balance the analytical requirements and the available resources. In studies involving attitude and descriptive scales, a reliability coefficient equal to 0.70 may well suffice, although the error variance is then in the order of 50 percent. In other studies involving the cross-national comparison of outcome measures, a reliability coefficient equal to 0.80 may be regarded as a minimum standard. In data collection involving individual estimates, the reliability requirements may well need to exceed the 0.90 level (see also

Chapters 6 and 8). If the respective reliability standards are not fulfilled, the derivation of composite measures may improve the situation.

Standards for Validity

Each measure should be tested for its validity with respect to the intended purpose. Standards for that purpose can often be established through a review of the literature that evaluates the validity of the data collection instruments; through an assessment of the variables by experts, data providers, and users; or through an assessment that determines whether the variable discriminates known differences and is consistent with results from related measures. Validity depends highly on the type of questions asked. For example, if data are collected on the extent of enrollment in higher education—where higher education can refer to different levels of educational attainment in different countries—such data can often not be used adequately for comparisons that make inferences about the quality or the level of achievement in higher education.

Standards for the Handling of Missing Data

Frequently, data collections fail to obtain data for particular questions and/or data categories. The instances of missing data that may be encountered in a particular data collection activity will, of course, depend on the design of the data collection activity. Data can be missing for a variety of reasons. First, there may be no data because a particular question is generically not applicable in a particular country, that is, the required data do not even potentially exist. For example, a question may refer to a certain level of education, a certain school type, or a certain type of expenditure which does not exist in a particular education system. Second, certain data may not be available in a country or may be available only in a form where the data are included in another category of the question or in another question of the questionnaire. This often occurs in situations where data are not available for certain sub-categories but can be provided for certain subtotals or totals. A third case is that no hard data are available but that the magnitude of the element is known to be negligible. In such a case terms like "negligible" need, of course, to be defined. Finally, it is possible that there may be no information for certain countries, tables, questions, or categories due to nonresponse.

The data producers usually assign special codes to missing data elements. The different reasons for the lack of data need to be distinguished so that they can be properly accounted for in the statistical analyses which will be undertaken with the data. It is therefore important that standards for the coding and handling of missing data are established. Such standards must ensure that the codes for missing data represent the different instances of missing data exhaustively, that the missing codes are mutually exclusive, and that there are clear definitions and instructions on how to assign the missing codes.

Standards for the coding of missing data must be supplemented by standards for the analytic treatment of missing data and their representation in the reports. For example, if data for a particular question or category are included in a category of another question, then the issue arises how these data are handled in the coding and analysis, especially how the missing data is related to the respective target category in which the data are included. If there are no standards, then this can lead to double counting and other inconsistencies. For example, if a country cannot provide separate data for educational expenditures in preprimary and primary education, and data for both levels of education are therefore included in the figures for primary education, then the statistics for primary education will be biased if this fact is not correspondingly dealt with.

For calculations, standards are further required that specify in each situation whether missing data are excluded from the calculation or substituted by an estimate. Countries may not have precise data available for certain classification categories of certain variables but they may still be capable of providing an estimate or a provisional substitute for these data. For example a country may collect no data on a variable but can create an estimate based on assumed relationships to other variables, or a country may not have data on the desired level of aggregation but may be able to provide aggregate figures for the desired categories on the basis of assumed relationships to other variables (for example, data may not be available at the national level, but can be aggregated from state or provincial figures). A further example involves the case where data may be available only for certain sub-populations but estimates can be provided for the remainder (for example, certain data may be available for public schools and government-dependent private schools but not for independent private schools). Finally, it may occur that the data are not available for the year of the data collection but it may be possible to provide estimates based on the data collected in previous years.

It is therefore important that the data requesters establish clear standards indicating how such situations should be dealt with by the data producers and the data providers. Solutions for the handling of missing data are usually very complex, especially when calculations involve multiple variables and multiple classification categories. If data are missing for a particular classification category, then this may mean that the values of the associated indicator cannot be calculated for that country or, alternatively, that assumptions about the missing data have to be made. Such assumptions carry a great potential for error, especially if they are applied centrally without taking into account the supplementary information that may be available in the country. For example, if there are independent private schools in a country but there are no data for these, and therefore the data for private schools include only the data for public schools and government-dependent private schools, then the data for that country will be biased, and there is no way of knowing by how much. If, on the other hand, the country can provide an estimate for the independent private schools, possibly with a lower and upper bound of the error, then corresponding adjustments can be made in the international comparative analyses.

In the case of estimated data it is important that standards exist which define and document the general problems associated with a specific variable or category; the accuracy of estimates; the kind of estimations undertaken and the estimation procedures used; the methods of calculation undertaken if the values are aggregates or derived from other variables; and national deviations from the definition of the variables or from their operationalization. Without such standards, different countries could treat estimations differently and this threatens the comparability of the results.

Standards for the Design and Operation of Data Collection

Equally important as standards for the results of data collections are standards concerned with the data collection operations themselves. These can ensure that the data collection operations are undertaken under comparable conditions; that the data preparation yields sound and comparable data; and that the data analysis and reporting are undertaken in accordance with the design and constraints of the data collection activity. Failures in each of these aspects can put the credibility of the entire data collection activity into question.

Standards for the Data Collection Design

Once agreement is reached on the aspects of education systems that are to be measured, these need to be defined in order to secure a general understanding of their meaning. The design will usually commence with the articulation of the policy and research questions that operationalize the information needs and decompose the issues investigated into their constituent parts. This requires the following steps. First, the definition of the constructs required for the reporting. Second, the definition of the variables which measure these constructs. Third, the definition of an internationally valid and consistent set of classification categories which operationalize these variables (such as "levels of education", "types of service providers", and "types of expenditures"). Lastly, it requires the specification of the type of data to be collected and the identification of appropriate data collection methods and adequate data sources.

When the data requirements are validated against these criteria, it sometimes turns out that these cannot be fulfilled under the given constraints. There can be several reasons for this. For example, it may not be possible to measure certain constructs adequately, or the costs in time and money for developing certain measures may be prohibitively high. Conceptual differences can also complicate matters, since they not only reduce the comparability of the data but also the possibility of collecting them. Further, it is frequently the case that standard definitions cannot be established for either political, methodological or other reasons, and hence, that the available data lack uniformity, reliability, coherence, and continuity.

The design must also establish whether the data needs are of a cross-sectional nature or whether there is a need to measure change over time. In the latter case additional standards will be required so as to ensure the reliability and validity of the data over time.

Based on the conceptual framework, internationally valid data collection instruments must be prepared and standards must be specified for the data collection instruments. These should establish precise and unambiguous instructions for their operationalization and measurement. They must ensure that the questions have the same meaning to all data providers and promote accurate responses. This, in turn, requires that the questions are presented using appropriate media and in a format that minimizes response errors, coding errors, and keying errors. Finally, the burden on the data providers in assembling and providing the data should be kept within acceptable limits.

Standards for Data Collection Operations

The methodology that is required for the collection of the data can range from aggregated summary statistics at the national level to outcome data obtained through complex surveys based on national samples of students, classrooms, and schools. Data quality standards need to be defined and procedures must be established which ensure that the data correspond to these requirements. All variables and operations must then be piloted under the same conditions that will be encountered in the data collection and with a size that ensures that the reliability of the results is high enough to allow the verification of the instruments.

A key decision is required whether data are collected with reference to the entire population in the form of a census or through probability samples in which every element of a defined target population has a known and non-zero probability of selection as expected in probability sampling. Many standards related to the data collection operations depend on this decision. Whereas certain questions require a census, others can profit from the advantages of a sample.

Sample survey designs generally offer the following advantages. They are usually much cheaper than censuses—in large populations, results that are accurate enough to satisfy the quality requirements of the survey can often be obtained from only a small fraction of the population. Because of this size reduction, the fieldwork can often be more rapidly conducted. Sample surveys can—at least potentially—increase the quality of the data, since the size reduction permits to divert resources into improving the quality aspects of the study. Sample surveys also allow a wider scope of data collection, because the use of more complex methods can often be only undertaken with a limited sample.

Further standards are required that can help to measure and control the various sources of error, for example, errors arising from the use of a particular method; errors related to noncoverage, unit nonresponse, item nonresponse, and sampling; or errors associated with the data collection instruments, methods, operations, and management.

Standards for Data Management

In order to obtain data that fit the reporting requirements and are of adequate quality, data management issues need to be considered in all phases of a data collection activity. This is particularly important for international comparisons in which the data collection and the data management are coordinated in different countries, possibly using different methods and operations. Here international standards must ensure the adequacy and comparability of the data and that the analytic demands on the data can be met. This involves several steps, such as the analysis of potential problems at an early stage in the data collection, the specification of quality demands and data cleaning rules; sufficient training of field administrators in the adherence to standards, the implementation of procedures to ensure that the collected data adhere to the standards specified; and the development of procedures for resolving observed deviations from the standards.

The careful preparation of administrative procedures including manuals and survey tracking instruments are important for the verification of the data quality standards. Also, the organization of the data into adequate data structures is required in order to manipulate, analyze, and report the information collected in a convenient and efficient way. Various technologies are available for these purposes and it must be decided which are the most appropriate within the respective administrative, logistical, and economic constraints.

The type of errors that can occur in the data management and which need to be accounted for through the establishment of corresponding standards, depend mainly on the instrument design, coding procedures, and the data collection and data entry methods. For example, the type of errors that arise when free response data are manually coded and then transcribed into computer readable form differ from the type of errors that are likely to occur when data are directly transcribed by computers from machine-readable answer sheets. For some of the problems it is nearly impossible to verify whether the errors have been caused by the respondent, during the field administration, the coding process, or during the data transcription. Yet many kinds of data analyses require that key decision rules be established to treat these problems, even if they cannot be resolved. The following are among the most common potential problems that must be addressed in the definition of data management standards:

1. The respondents may have been assigned an invalid or wrong identification code either during instrument preparation, field administration or data transcription;

2. Questions may have accidentally been misprinted due to technical or organizational imperfections, thereby preventing respondents from giving an appropriate answer;

3. Questions may have been skipped or not reached by the respondents either in a randomized fashion or in a systematic way, or respondents

may have given multiple responses when only one answer was allowed, or questions may have been answered in other unintended ways;

4. The data values may not correspond to the coding specifications or the range validation criteria specified;

5. Answers to open-ended questions may contain outlier codes, that is, values which are improbably low or high even though they could be the true answers;

6. The data from a respondent may contain inconsistent values, that is, the values for two or more variables may not be in accord; and

7. Inconsistencies between data values from different respondents, or groups of respondents, may occur.

Procedures are required for checking invalid, incorrect and inconsistent data, which may range from simple deterministic univariate range checks to multivariate contingency tests between different variables and different respondents. The criteria on which these checks are based depend on the type of data and on the manner and sequence in which the questions are asked. For example, for some questions a certain number of responses are required, or responses must be given in a special way, due to a dependency or logical relationship between the questions.

Depending on the data collection procedures used, it must be defined when and at what stage data management standards are implemented. For example, some range validation checks may be applied during data entry, whereas more complex checks for the internal consistency of data or for outliers may be more appropriately applied after the completion of the data entry. Problems that have been detected through the verification procedures need to be resolved efficiently and quickly. Some problems can be resolved automatically through a cross-validation with responses to other questions or with responses from other respondents. Standards are required that ultimately allow to answer the following questions for each problem: When to correct a data value on the basis of other data values? When to set a data value to "missing"? When to drop the data for a respondent because of invalid, missing or not administered data? And finally, when to drop a question or variable?

Standards for Data Analysis and Reporting

The data analysis and reporting stages are the final step in a data collection activity. This step synthesizes the design, data collection, and data management stages in order to answer the intended policy and research questions. It documents the substantive study findings and the technical operations and analysis procedures. The methodology that is used depends highly on the particular aims and design of the data collection activity, and correspondingly also the standards that are required in this phase vary.

Based on the standards for the results and on the questions to be answered, an analysis plan needs to be prepared which should specify: the general form of the comparisons that are undertaken to answer the intended questions; the analytic techniques employed; the entities for which data must be collected and the entities for which data will be reported; the survey variables involved and their operationalizations and interrelationships; and the specific statistical analyses that are required.

Such an analysis plan has usually a dynamic nature and serves different purposes in different phases of a data collection activity. An initial analysis plan can help to verify that the design of the data collection covers all of the intended questions and that the questions, the methodology, and the data collection instruments are adequate and internally consistent. Later on it can provide a basis for the preparation of the data analysis procedures, and once data are available it can guide and document the actual data analyses.

Sometimes, large differences between countries can be due to sampling errors and it is therefore necessary to establish standards for the statistical significance of detected differences. It is moreover important that the specific hypotheses are defined prior to the data analysis; that confidence intervals are calculated and reported for all key statistics; that significance levels for the results of statistical tests are specified before inferences are made and reported; and that the levels of significance conform to the substantive requirements. For significance tests involving multiple hypotheses, correspondingly multiple comparison procedures must be used, especially when confidence intervals are established for statistics that are compared across countries.

An important aspect is to provide the ultimate users of the published data with sufficient information so that they can evaluate the quality of the data and the results. Standards should ensure that information is provided on the sample sizes and response rates, the reliability and validity of the results, the sampling errors for all statistical estimates used; error rates of data transcription and an indication of data cleaning problems encountered; an indication of the major sources of non-sampling errors and the rules that were applied when dealing with missing and inconsistent data. Finally, information on the distributions of the variables should also be provided.

Conclusion

The purpose of international and national indicator systems is often to allow comparisons—of countries which one another, of a single country's progress over time, and of the progress of several countries over time and in comparison with each other. Since indicators are intended to allow valid comparisons over time and across countries, their usefulness obviously depends on the reliability, validity, and comparability of the data. To attain this comparative purpose more fully, a greater degree of similarity in the design and operation of data collection procedures, variable specification, measurement, data treatment, and data analysis and reporting must be secured. This chapter has dealt with several aspects of the teachnical standards and criteria that are needed for international education statistics and indicators. It has explained what standards are needed for what areas, how they may be derived, and how they can be used to secure a general and internationally valid understanding of the comparison.

References

Cooperative Education Data Collection and Reporting Standards Project (CEDCAR) Task Force. (1991). *Standards for education data collection and reporting.* Washington, DC: National Center for Education Statistics, US Department of Education.

Elley, W.B. (1994). *The IEA international study of reading literacy.* Oxford: Pergamon Press.

Foreman, E.K. (1991). Survey sampling principles. In D.B. Owen (Ed.), *Statistics: textbooks and monographs.* New York: Marcel Dekker.

Groves, R.M. (1989). *Survey errors and survey costs.* New York: Wiley.

Joint Committee on Standards for Educational Evaluation. (1981). *Standards for evaluations of educational programs, projects, and materials.* New York: McGraw-Hill.

Kalton, G. (1983). *Compensating for missing survey data.* Institute for Social Research, University of Michigan.

Kish, L. (1965). *Survey sampling.* New York: Wiley.

Livingston, S.A. (1988). Reliability of test results. In J.P. Keeves (Ed.), *Educational research, methodology, and measurement: An international handbook.* (pp. 386-392). Oxford: Pergamon Press.

OECD (1993). *Education at a glance II.* Paris: Organization for Economic Cooperation and Development, Center for Educational Research and Innovation.

Ross, K.N. (1987). Sample design. *Evaluation in Education 11,* 57-78.

Ross, K.N. (1988). Sampling errors. In J.P. Keeves (Ed.), *Educational research, methodology, and measurement: An international handbook* (pp. 537-541). Oxford: Pergamon Press.

Rossi, P.H. (Ed.). (1982). *Standards for evaluation practice.* San Francisco: Jossey-Bass.

Schleicher, A., & Umar, J. (1992). Data management in educational survey research. *Prospects 83,* 317-325.

Tuijnman, A.C. (1994). Selection bias in educational research. In T. Husén & T.N. Postlethwaite (Eds.), *The international encyclopedia of education, 2nd edition.* (pp. 5379-5385). Oxford: Pergamon Press.

About the Authors

Albert E. Beaton, Ed.D., is a Professor at Boston College. He is currently the Director of the Center for the Study of Testing, Evaluation and Educational Policy at Boston College, as well as Study Director for the IEA Third International Mathematics and Science Study. His teaching responsibilities include courses on technical issues in educational testing and multivariate statistics, among others. Mailing address: Boston College, Campion Hall 323, Chestnut Hill, MA 02167, USA.

Norberto Bottani is Head of the Unit for Education Statistics and Indicators at the Organization for Economic Cooperation and Development. Prior to coming to this organisation he served in the Swiss Federal Administration, where he was involved in the formulation of educational research policy for Switzerland. While at the OECD he led a major study on early childhood education and intersectorial policies for child development. He was also in charge of a project on education and cultural and linguistic pluralism. He has led OECD's effort to develop a limited set of international education indicators since the work began in 1987, and is author or co-author of several books. Among these are *Education at a Glance I & II* (OECD, 1992 & 1993), and *Making Education Count: Developing and Using International Indicators* (OECD, 1994). Mailing address: OECD, Center for Educational Research and Innovation, 2 rue André-Pascal, 75775 Paris 16, France.

Mary Lyn Bourque is currently on the staff of the National Assessment Governing Board in Washington, DC as the Assistant Director for Psychometrics. She serves as the Board's chief technical staff on all matters related to the design and methodology of NAEP assessments. Prior to taking up this position, Dr. Bourque spent 15 years teaching secondary-level science and mathematics before becoming the Director of Testing in a large urban city in New England. She has also taught in the graduate Department of Measurement and Statistics at the University of Maryland. Her area of interest is applied measurement, and she has published extensively in that field in the form of journal articles and technical reports. Mailing address: National Assessment Governing Board, 800 North Capitol Street, NW, Suite 825, Mailstop 7583, Washington, DC, 20002-4233, USA.

Vernon C. Daniel is a doctoral candidate in Public Policy Analysis at the University of Illinois at Chicago. He was awarded a Master's degree in Comparative Education at the Florida International University; a B.Sc.

degree in Vocational, Technical, and Adult Education at the University of Wisconsin; a professional certificate in education at the University of Birmingham, United Kingdom, and a technical teachers' diploma at the Wolverhampton Technical Teachers College. He served as a curriculum development officer in the Ministry of Education, Republic of Trinidad and Tobago, and was awarded a Commonwealth Bursary by the government of the United Kingdom and a fellowship by the US government under the Caribbean Basin Initiative. Daniel is interested in issues of quality control, effectiveness and productivity as they relate to the development and organization of management systems in education in both the developing and developed world. Mailing address: University of Illinois at Chicago, 522 North Euclid Avenue, Oak Park, IL 60302, USA.

Eugenio J. Gonzalez is currently completing his doctoral work in Educational Measurement and Evaluation at Boston College. He is also a research assistant at the Center for the Study of Testing, Evaluation and Educational Policy at Boston College, where he provides statistical and psychometical assistance to several ongoing projects, including the IEA Third International Mathematics and Science Study. Mailing address: Boston College, Campion Hall 323, Chestnut Hill, MA 02167, USA.

Torsten Husén was born in Lund, Sweden in 1916. He is a Professor Emeritus at Stockholm University since 1982. Before that he held numerous academic positions in Sweden and abroad. He was chairman of the International Association for the Evaluation of Educational Achievement (1962-78), and served as chair or trustee on the boards of the International Institute of Educational Planning (1970-80), International Council for Educational Development (1971-present), and the International Academy of Education (1986-present). With T.N. Postlethwaite he was the Editor-in-Chief of the International Encyclopedia of Education (1st edn 1985, 2nd edn 1994). He is the author of some 50 books and over 1000 papers and articles listed in *Torsten Husén: Printed Publications* (Stockholm, 1981) and *Torsten Husén: Publications Supplement 1981-91* (Stockholm, 1991). Mailing address: Institute of International Education, Stockholm University, 106 91 Stockholm, Sweden.

Ina V.S. Mullis, Ph.D., is Executive Director of the US National Assessment of Educational Progress. Her interest and fields of specialization are large-scale assessments, performance assessments, and education indicators. Mailing address: Educational Testing Service, National Assessment of Educational Progress, Princeton, New Jersey, 08541-6710, USA.

R. Murray Thomas was Head of the Program in International Education at the Graduate School of Education at the University of California, Santa Barbara, where he has held a professorship since 1961.

He earned a Ph.D. at Stanford University in 1950 in Educational Psychology with emphases on child development and counseling. His publications include more than 150 articles and research reports as well as some 30 books dealing with a variety of topics in child psychology and comparative education. Mailing address: 1436 Las Encinas Drive, Los Osos, CA 93402, USA.

Eugene H. Owen is Chief of the International Activities Group, Data Development Division, National Center for Education Statistics. He is specialized in international education statistics and large-scale assessments. Mailing address: United States Department of Education, National Center for Education Statistics, 555 New Jersey Avenue, NW, Washington, DC, 20208, USA.

Gary W. Phillips, Ph.D., is Associate Commissioner of the Education Assessment Division at the National Center for Education Statistics. He specializes in statistics and psychometrics. Mailing address: National Center for Education Statistics, US Department of Education, 555 New Jersey Avenue, NW, Washington, DC, 20208, USA.

T. Neville Postlethwaite is a Professor of Comparative Education at the University of Hamburg since 1976. He received his first degree in Psychology from the University of Durham. He earned a Ph.D. in Educational Psychology at Stockholm University in 1967, after having been a school teacher in England, a researcher at the National Foundation for Educational Research in England and Wales, and Executive Director of the International Association for the Evaluation of Educational Achievement (IEA). He has undertaken many comparative studies in some 40 countries across the world. Most of these have focused on the prediction of cognitive achievement. His most recent publications dealt with the prediction of science achievement in 23 countries (with David Wiley, 1992), effective schools in reading in 32 education systems (with Kenneth N. Ross, 1993), and a 12-volume *International Encyclopedia of Education* (edited with Torsten Husén, 1994). Mailing address: Institute for Comparative Education, University of Hamburg, Sedanstrasse 19, 20146 Hamburg, Germany.

Andreas Schleicher, M.Sc., is Director of Data Management and Analysis for all ongoing studies of the International Association for the Evaluation of Educational Achievement (IEA). Before taking up this position he was a project manager at Phillips Medical Systems and International Coordinator for the IEA Reading Literacy Study. He is currently also a consultant to OECD's education indicators project and acts as a resource person for World Bank and UNESCO projects aimed at the improvement of international and comparative education statistics. Mailing address: IEA Secretariat, Sweelinckplein 14, 2517 GK, The Hague, Netherlands.

Judith Torney-Purta is Professor of Human Development at the University of Maryland at College Park. She earned an A.B. in psychology (with honors) at Stanford University and an M.A. and Ph.D. in human development from the University of Chicago. As a developmental psychologist she has specialized in the study of political socialization, civic education, and social cognition in the United States and cross-nationally. She is a fellow of the American Psychological Society and the American Psychological Association. She served as a member of the Board on International Comparative Studies in Education of the National Academy of Sciences in the US from its inception in 1988 through 1994. Mailing address: University of Maryland at College Park, Institute for Child Study, College of Education, 3304 Benjamin Building, College Park, Maryland, 20742-1131, USA.

Albert Tuijnman was born near Maastricht in 1959. He was awarded a Ph.D. in Comparative Education at Stockholm University in 1989 and is currently staff member of the Center for Educational Research and Innovation at the OECD in Paris. He is author of over 75 journal articles and book chapters, and has written and edited 12 books. The most recent ones are: *Schooling in Modern European Society: A Report of the Academia Europaea* (Pergamon, 1992); *Learning Across the Lifespan* (Pergamon, 1992); *Effectiveness Research into Continuing Education* (Pergamon, 1992); *Education at a Glance I* (OECD, 1992); *Education at a Glance II* (OECD, 1993); *Making Education Count: Developing and Using International Indicators* (OECD, 1994); and *Educational Research and Reform: An International Perspective* (US Department of Education, 1994). Mailing address: OECD/CERI, 2 rue André-Pascal, 75775 Paris 16, France.

Herbert J. Walberg was awarded a Ph.D. in Educational Psychology by the University of Chicago in 1964. He held research appointments at that university, the Educational Testing Service and the University of Wisconsin, and taught at Harvard University. In 1984, Walberg was appointed Research Professor of Education at the University of Illinois at Chicago. His main research interest is the educational promotion of human learning and talent. He has written or edited 52 books on various topics and wrote 67 chapters, four encyclopedia articles, 23 pamphlets and technical monographs, and about 300 research papers. In recent years he was elected a Fellow of the American Association for the Advancement of Science, a Fellow of the Royal Statistical Society (London), a Fellow of the American Psychological Association, and a Visiting Fellow of the Australian Association for Educational Research. With Benjamin S. Bloom, James S. Coleman, Torsten Husén, and T. Neville Postlethwaite he served as a founding member of the International Academy of Education. Mailing address: University of Illinois at Chicago, 522 North Euclid Avenue, Oak Park, IL 60302, USA.

Richard M. Wolf is a Professor of Psychology and Education and Chair of the Department of Measurement, Evaluation, and Statistics at

Teachers College, Columbia University in New York. Previously he taught at the University of Southern California and the University of Chicago. Professor Wolf has published many articles in the areas of measurement and evaluation. His books include *Evaluation in Education* (Third Edition, 1990), *Achievement in America* (1977), and *Current Issues in Testing* (with Ralph Tyler, 1974). Mailing address: Box 165, Teachers College, Columbia University, New York, NY 10027, USA

Guoxiong Zhang received a B.A. from East China Normal University in 1988 and a Master's degree in Education from the University of Illinois at Chicago in 1991. Since then, Zhang has been a consultant and data analyst for the Department of Research, Evaluation, and Planning in the Chicago Office of Public Schools, concentrating on the evaluation of programs for gifted students. Zhang is completing his doctoral studies in educational evaluation at the University of Illinois at Chicago. His interests are in research methods, policy studies, international comparative studies, and the effects of environment and personality on school success. Mailing address: University of Illinois at Chicago, 522 North Euclid Avenue, Oak Park, IL 60302, USA.

Author Index

Subject Index